HOME GOAT KEEPING

The *Invest in Living* Series
Home Goat Keeping
Home Poultry Keeping
Home Rabbit Keeping

INVEST IN LIVING

HOME GOAT KEEPING

by

L.U. HETHERINGTON

A & C Black · London

First edition 1977
Reprinted 1979, 1981, 1983, 1986

Published by A & C Black (Publishers) Limited
35 Bedford Row, London WC1R 4JH

Text and photographs copyright © L U Hetherington 1977, 1979, 1983, 1986

Hetherington, L.U.
   Home goat keeping.—(Invest in living)
   1. Goats
   I. Title   II. Series
   636.3'9083   SF383

   ISBN 0–7136–2690–9

ISBN 0 7136 2690 9

Printed and bound by The Garden City Press Ltd, Letchworth, Hertfordshire SG6 1JS.

Contents

Why Keep Goats?   7
Health for you

What to Buy   9
Toggenburg
British Toggenburg
British Alpine
Saanen
British Saanen
Golden Guernsey
English Guernsey
Anglo-Nubian
British

Where to Buy   13
Markets and goat clubs
Age
Good goats

Housing   16
Siting the house
Suitable buildings
Pens
Food racks
Outdoor fencing
Tethering

Feeding   25
Hay, pea and bean
Greens, roots and fruit
Concentrates
Analysis of various acceptable foods
Sample rations
Newly kidded
Young goats
Growing food
Making hay
Wild plants
Poisonous plants

Breeding   34

Kidding   37
Abnormal kiddings
Kid rearing
Sexing kids

The milk
Feeding kids
Horns

Milking   44

Uses of Milk   45
Clotted cream
Butter
Cheese
Yoghurt
Other products

Milk Recording   50

Showing   53

General Care of Goats   55
Keeping a male
Management
Other kinds of stock
Disposal of kids
Some do's and don't's

Ailments   61
General precautions
Drenching
Abortion
Abscesses
Acetonemia
Anaemia
Arthritis
Blindness
Bloat
Caprine arthritis and encephalitis
Coccidiosis
Colic
Cuts
Entero-toxemia
Fading
Fluke
Footrot
Fractures
Grass tetany
Laminitis

Lice
Lumpy jaw
Mastitis
Metritis
Milk fever
Pink milk
Pneumonia
Poisons
Pulpy kidney
Rheumatism
Rickets

Scurf
Tetanus

Glossary   68

Sources of Further Information   69
The British Goat Society
Suppliers

Index   71

About the Author

Mrs Hetherington started keeping goats for milk during the war when her son developed a glandular infection from cow's milk, and ever since she has bred them in Suffolk. From 1964 she has served on the committee of the British Goat Society, and for several years was their Public Relations Officer. She has exported goats to countries in many parts of the world, including the Sultanate of Amman, the West Indies and West Africa. One of her Toggenburgs, Alderkarr Damaris, three times recorded the highest Toggenburg milk yield in the British Isles, and many of her goats have won prizes at British Shows.
In 1978 and 1980, the author's Toggenburg Q*[1] R166 K Alderkarr Davinah was the highest recorded Togg. She is a daughter of Damaris. In 1979, *R152 K Alderkarr Dryope (granddaughter of Damaris and niece of Davinah) was top Toggenburg. Also in 1979, Alderkarr Genista R120 was the only Golden Guernsey to get her recording R, so Davinah's yield equalled 3700 lbs + for the year.

The author's Alderkarr Carradoc was the first male Togg to become a Sire of Merit; he was recorded as SM.Br Ch 826/35. The latest Toggenburg male to be added to the list of 8 is Alderkarr Dane, SM.Br Ch 166/152.

Why Keep Goats?

Goatkeeping should definitely be considered by those who are interested in self-sufficiency. It cannot be overestimated as an interesting hobby which will provide the family with a supply of fresh, untreated dairy produce.

Those who think that goats have an unpleasant smell may rest assured. Female goats do not, and no one would suggest your keeping a male, which, in the rutting season, does smell.

Costs, like everything, are rising, but feeding a goat should cost no more than feeding a large dog, and no dog will give you, say, six pints of milk daily in summer and about three pints in the winter.

It should be borne in mind, however, that milking must be done twice daily throughout the year; if you are prepared to do this, go ahead.

Holiday: If you are in the habit of going away for a time each year, provision for a replacement milker must be made; or else you must arrange to take the goats to some other goatkeeper who will care for them during your absence. It is always a good thing for more than one person in a household to be able to milk, in case the usual milker should have an illness, however temporary. My own children were all taught to milk at six years of age.

Keen gardeners and allotment holders can grow a good portion of the bulk food loved by goats. The animals will also use up excess vegetables and clean kitchen trimmings. The very best reason of all for keeping goats is because you like them.

Health for You

Many goats give a gallon of milk, and some over two, but the household milker which gives six to seven pints a day in summer, and three to four in winter, is about enough for most people. The sensible number of goats to keep is two, a sort of 'belt and braces' operation. It means that, kidding one each year, you always have one milking whilst the other is dry, prior to kidding. It gives you quite a lot of milk in the summer, and enough in the winter.

One only has to spend an afternoon on the information stand at any agricultural show to realise just how many country children in the past were sickly until they were reared on goats' milk: 'Our John always had goats' milk. Proper poorly he was till then, but he got on wonderful after that.' 'My dad always used to keep a goat on his allotment,' etc. And there never was any goat kept that, according to its owners, did not give at least that magical amount of a gallon a day.

Both children and adults who suffer from eczema, asthma or psoriasis may be allergic to beef protein and benefit by changing from cows' milk to goats'. Again, those suffering from stomach disorders should find goats' milk much easier to digest as it has a smaller amount of caesin and a greater amount of Vitamin A[1]. The fat globules in goats' milk are very small and stay mixed with the milk instead of rising as a cream, so that a soft curd develops in the stomach in twenty minutes, as against the two hours required to curd cows' milk. Really, young infants do very well on goats' milk, although, like all milk feeds, it is short on iron. Anyone fed solely on any type of milk, cows' or goats', would suffer from anaemia unless their diet was supplemented by some form of iron, as for example in rose hip syrup.

Lucky is the child of a goatkeeper who gets his fresh supply of goats' milk, cheese, yoghurt and cream. It seems that in Asia Minor they live to 180 years old on goat yoghurt; I should not like to do so myself, but there is no accounting for tastes.

What to Buy

There are eight recognised breeds of goats in the British Isles; they are as follows:

Toggenburg. A direct descendant from Swiss imported goats, with no outcrossing, this is a small pale fawn goat, with a longish coat, a white stripe on its face and white legs and rump. It is very pleasant-natured, frequently hornless, a good steady milker of long lactation and a good grazer.

British Toggenburg. This is much like the Toggenburg from which it was bred, but larger, darker in colour and short haired. It is a good milker with long lactation. Both these Toggenburg types, and most Swiss (short prick-eared types), sometimes have small appendages on the neck, called tassels. These appear to have no real use now, apart from protecting the thyroid gland from cold weather, and as an adornment.

British Alpine. Much like the British Toggenburg, but black with white markings and of very smart appearance, this is a large rangy animal which needs plenty of exercise and is therefore unsuitable for total stall feeding.

Toggenburgs: Alderkarr Hebe and her two-month-old daughter, Alderkarr Hepsibah, bred and owned by the author. At the Suffolk Show, June 1976, they won the Best Mother and Daughter award

British Alpine goatling, Tamar Briarose, a winner at the Royal Norfolk Show

A Golden Guernsey goat with her three-day-old kid

Saanen Mostyn Merrymiss

British Toggenburg goatling

An Anglo-Nubian first kidder

The author with a selection of her goats of all breeds

Saanen. Again from direct Swiss imported stock, this docile animal has a short white coat and is an excellent milker.

British Saanen. This is much like the pure strain, but larger. It is probably the heaviest milker of all, will stall feed quite well, but also grazes freely.

Golden Guernsey. The small golden goat of Guernsey in the Channel Islands, this breed was almost extinct until a few years ago when it was made into a Trust in the island and some animals were brought to the mainland for breeding. It is now a rare breed, and growing in number and popularity. The animals are very small, and as they are docile and much attached to their owners, are excellent for young or older folk. They give a good yield for size, are good grazers and have proved adaptable to stall and yard feeding if started young.

English Guernsey. Breed added to bring up numbers and help remove the 'in breeding' of Golden Guernsey. Procured by mating a Golden Guernsey female with a Saanen or British Saanen male; female progeny go into a SR (G) register. Their female progeny, by another Golden Guernsey male, go into the IR (G) register, and female kids from IR (G) dams, again by Golden Guernsey males, are then registered as English Guernsey. Males may be registered also from this last mating, as English Guernsey; from then on kids by Golden Guernsey or English Guernsey males are eligible for English Guernsey register, but neither males nor females can ever enter the Golden Guernsey section.

Anglo-Nubian. The goat of the desert and Bible, it is large, has long drooping ears and a Roman nose and comes in all colours—mixed, spots and marbling being frequent. As it has the highest butter fats, it is known as the Jersey of goats, though it is not such a long lactation type as the Swiss breeds. It is a very attractive animal.

British. This is not really a breed, but the result of crossing two pedigree animals to obtain some particular point. The goat is usually an excellent producer, and as it can resemble either parent, it can appear to be of any given breed.

The British Goat Society, which aims to maintain the welfare of goats, also has several Grading Registers. Progeny from unregistered dams can eventually be brought up to the Breed Register by the use of good males. This does not apply, however, to Toggenburg, Saanen and Golden Guernsey; these are pure strain, obtainable only from two of their own breed.

There are now specialist Breed Societies for many breeds, all affiliated to the BGS. They include Golden Guernsey Goat Society, Anglo-Nubian Breed Society, Toggenburg Breeders Society and also Societies for British Alpines, British Toggenburgs, Saanens, and the latest is the hoped-for revival of the English Goat, thought to have been extinct, but some specimens have been found, well spread, and are now being conserved by the English Goat Society. All addresses may be obtained from the BGS.

Where to Buy

Markets and Goat Clubs

Where to buy is really quite a problem for beginners. The first thing to remember is not to buy on markets or from dealers who advertise everything from a mouse to an elephant. It is the quickest way to frustration and disappointment.

Animals in markets are there, as a rule, because in some way they are deficient. They may possibly be animals which give a fair supply of milk immediately after kidding, but go dry very fast, or they may have some other drawback which is not immediately visible. These are the goats bought by a dealer and then passed off as something other than what they are to fetch a higher price; they are not a good bargain. Of course, there are exceptions, and you might get a nice animal; but you are taking a considerable chance.

The British Goat Society, previously mentioned, has over fifty local goat clubs affiliated to it. These cover the British Isles and some areas overseas as well. The society will put you in touch with the club nearest to you, whose members can advise you on finding stock for sale from reliable sources. The price of a registered goat might be a little higher, though not always, and you might as well get the best for your money, time and care. The well-bred animal will have records which you can see, and will take no more food or time than a scrub goat.

Advertisements in local newspapers can be misleading, as folk advertise their animals as belonging to some specified breed, when they merely resemble that kind. Always request the goats' registration numbers and ask their parentage; if this information cannot be given, think again; and never buy an animal just because you are sorry for it.

Joining a local club, which will usually only cost you a small annual subscription, will bring you in touch with others interested in the same animals as yourself, and will enable you to register your kids through the club. The British Goat Society would also happily enrol you as a member, entitling you to a supply of interesting literature from them.

Age

The age at which to buy is a poser. Can you milk? If not, do not launch out into buying a milker unless someone is at hand to teach you, or you could very soon dry the goat off. If the animal is not 'stripped out' at each milking, she will consider that you do not want that amount and secrete less until she is dry; so learn to milk first.

Kids—well, they are charming creatures, playful and pretty. But they are a long-term project, needing four feeds of milk daily to begin with, and some milk for at least six months, plus other food and plenty of hay; time is involved too.

My personal choice would be to buy a goatling; she is either ready to serve in the autumn, or already served; you get to know her, and she, you. She is used to your feeding and her housing before she kids; you also have her kids and the whole of her lactating life before you.

Goats are herd animals which really hate being alone, so if you can possibly do so, buy two. These should not necessarily be of the same age group; if, for instance, you buy a goatling and kid, the kid will be ready to breed in her second autumn, and the goatling, then your milker, will milk through the winter. From then on, breeding one of them each year, you will never have a dry time.

Good Goats

What to look for: it is possible to look at goats at agricultural shows, where frequently a goat section is to be found. You can decide which kind you like, talk to their owners, and maybe buy or book something which is just what you want. Failing that, a book called *Breeds of Goats*, with photos of all available breeds included, can be obtained from the British Goat Society.

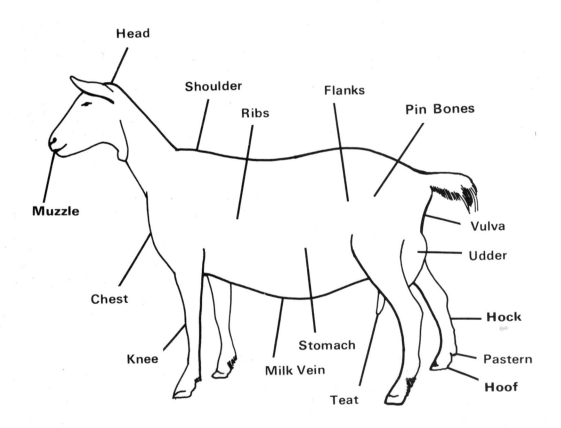

The points of a milking goat

Good goats should have a nice bloom on their coats, bright eyes, and be interested in all that goes on. Looked at from front to back, the animal should be triangular in shape, with slim shoulders and wider hips. There should be a straight line from neck to tail, with only a slight slope to tail, and the body should be deep, this depth increasing to the udder, which should be firmly attached to the body. The udder should be well attached forward and not narrow and necky; and remember that a long udder is liable to be damaged by brambles, etc., and can become sore and difficult to milk. The goat should have good feet, strong legs and a deep chest; a good spring of rib will show that she eats plenty of bulk. If you get all that, you should have a winner.

Housing

Siting the House

Before you begin keeping goats you should check with your local authority that the keeping of livestock does not infringe some bye-law or other. At the same time you should check whether any building you intend erecting is sufficiently large to need a building permit or planning permission. It will not have to be very big to do so. I had to get planning permission for my own goat house, and I had to fill in as many forms as if I wanted to build a tower block of flats. Imagine how you would feel if, having made a nice new house for your two milkers, you had to take it down again; it does not bear thinking about, does it?

Housing should always be prepared before you bring your goats home.

If you are putting up a new building for your goats, remember that a south-facing aspect is best. Access from a hard path is desirable for the delivery of foodstuffs, hay and straw. How close can you arrange your water supply? Water is heavy to carry, so consider the matter before anything permanent is done. If your goat stable is to be situated some distance from your water supply you will save a lot of time and effort if you stand a water butt or tank close to the stable so that you simply dip the pail in the water to fill it when you feed your goats. The tank can be refilled from time to time from your domestic supply by means of a hosepipe with a tap screw at one end. A good rain butt catching water off the roof will supply soft water for much of the year.

When you locate your goat house you must also consider the colder part of the year: you will not mind walking across the field on a lovely spring or summer morning, and it is positively pleasant to milk in the cool evening of a hot day. But it is not pleasant in wind and sleet or snow or when your fingers are freezing on to pail handles in the winter. So, within reason, situate it all as close to your own house as you can.

Suitable Buildings

Many existing buildings can be adapted for goats, with very little trouble.

Nissen huts, garages, large poultry houses, loose boxes and stables can all be used. As goats are smaller than most farm animals and, though inquisitive, are less destructive than such animals as pigs, lighter sheds can be utilised so long as they are draught-free. Sectional buildings designated by the manufacturers as 'deep litter poultry houses' make splendid goat houses. The dimensions vary, of course, but as a rule they are in sections of 8 ft. (2.4 m) and are made on good framing, clad either with flat asbestos sheeting or wood. Both materials are fine, although asbestos is best lined in some way, because it becomes brittle with age; it is, however, ratproof and fireproof. Width is either 16 ft. (4.8 m) or 20 ft. (6 m) and the length can be what you require in the section dimensions; ends

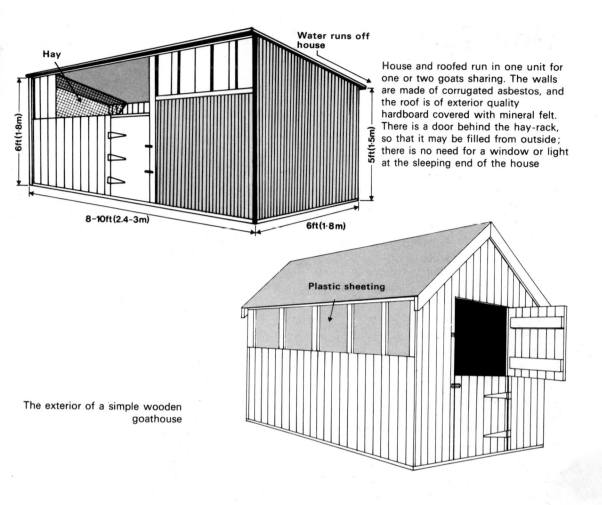

Hay

Water runs off house

6ft (1·8m)

8–10ft (2.4–3m)

6ft (1·8m)

5ft (1·5m)

House and roofed run in one unit for one or two goats sharing. The walls are made of corrugated asbestos, and the roof is of exterior quality hardboard covered with mineral felt. There is a door behind the hay-rack, so that it may be filled from outside; there is no need for a window or light at the sleeping end of the house

Plastic sheeting

The exterior of a simple wooden goathouse

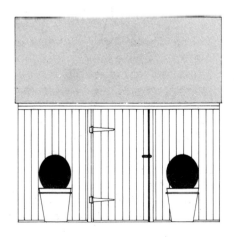

A garden shed 8 ft. (2.4 m) × 6 ft. (1.8 m) with a partition across 18 in. (450 mm) inside, this is suitable for two compatible goats or two goatlings. There are two openings for food pails

17

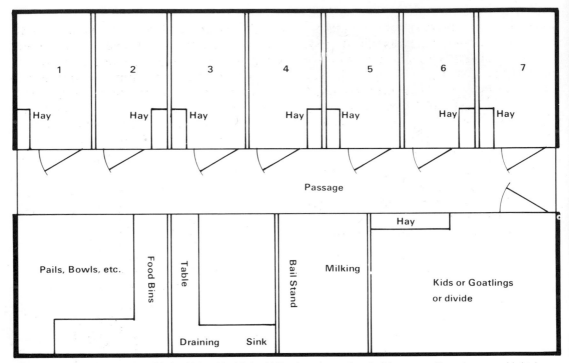

The layout of a Nissen hut 23 ft. (4.6 m) × 15 ft. (4.3 m) converted to hold seven milkers in pens. Light is provided by panels let into roof and ends

are supplied with doors on runners, but these are very draughty, and can be replaced with normal hinged doors in two sections.

If the walls of the building are merely a single layer of wood, a lining of some kind will make it warmer. If another sheeting of wood is out of the question, a great deal of difference can be made by wire netting and an insulating material. The netting should be stretched and nailed to the framing and the insulating material filled in between this and the wood.

To minimise heat loss (and hence food consumed) a false ceiling should be put in if the present ceiling in your goat building is high. The base can be made of wire netting affixed to battens. This should be covered with bituminous paper or even split-open meal bags—anything which will keep out draughts and stop condensation from dripping on to the back of the animals below. Do remember that a cold building will not only be cold and uncomfortable for your animals: you will also suffer yourself during the winter whilst you are milking and feeding in the chilly early morning and evening. More comfort for man or woman—and beast too—makes sense.

Goats have small hooves which soon cut up the yard earth so that it gets muddy in wet weather, with the result that the straw bedding is made dirty. If you have a small yard and no grazing area it is of great benefit to reduce the amount of mess by having both the yard and the floor of the house concreted, or paved with slabs.

If possible, drainage should be arranged to keep bedding dry. When concreting floors it is simple to make a drop of 2 in. per yard (approximately 51 mm per metre) to allow for the free removal of liquid. If your floors are earth, make a soakaway by locating the lowest part and digging a wide hole approximately 1 ft. 6 in. (457 mm) deep. This should be filled with gravel, covered again with earth and rammed hard.

Remember to arrange the internal structure so that you can get a barrow along any passage: cleaning out is a continual affair, and a good barrow makes life simpler.

Inside requirements for goats are:

- deep straw beds which are dry and out of draughts
- some light and air
- a rack for hay
- food and water pails
- a mineral lick (usually to be bought from the miller, along with the meal); this hangs by a cord on the wall.

A great deal of time and effort (especially in wet weather) will be saved if you can include space for some of the following items under the same roof as the goats:

- supplies of hay and straw
- a bin with a lid to keep food in and vermin and inquisitive goats out

Close by should be a place where dung can be stacked to rot down after the goats have been cleaned out. The resulting manure is excellent for the garden. To keep the heap within bounds and help it compress and heat quickly, it should be stacked neatly in the form of a box with straight sides, the shorter material being thrown into the centre. In urban areas where you cannot use the manure the drier parts of the stack should be burned in an incinerator.

It is of great advantage to ensure that there is a barrier to prevent your goats rushing out when you open the door. Otherwise you may find yourself dropping an armful of hay or spilling a pail of water—and this can be irritating! The best type of door is one which opens in top and bottom sections. Air and sunlight can enter through the open top section while the animal is restrained

Mineral licks

by the closed bottom half, although it may still be able to look over the top. It is now possible to buy hinges from which the doors can be lifted, they can then be tied to partitions making sure the animals don't injure themselves.

Yarded goats should be free to run in and out at will; they will know when they prefer to be out, but in any case should be in at night.

Goats lack the natural oil which most farm animals have in their coats, and consequently hate the wet. Those that have grazing, either in paddocks, herded or tethered, will suffer if they are left out in all weathers, as one sometimes does see. Common sense will tell you that it takes more food to warm up and restore a chilled animal than would be required if it was left in its house with a rack of hay on bad days.

Always use at least two strong T

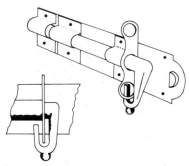

This kind of bolt, which has the movable part hooked at its handle instead of the usual knob, fools the goats because, as the hooked end turns back behind the ring through which the bolt passes, the animals cannot get hold of it with their mouths. It can also have a padlock threaded through it, which is an additional safeguard in areas in which vandalism is rife

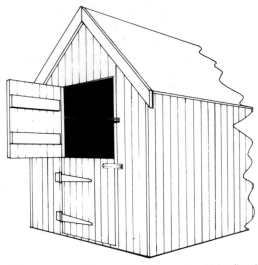

Wherever possible, stock buildings should be fitted with these stable-type doors. The upper half can be opened to allow air and light in, while the closed bottom half keeps out floor-level draughts. The door is held open by a wire fixed to two staples, one on the door and one on the outer wall

hinges to each door, because goats like to stand with their feet on top of the door or on a ledge if there is one. As an adult female is about 150 lb. (68 kg), you can imagine the strain on the hinges.

Bolts are the best fastening, but should not be less than 15 in. (381 mm) from the top of the door, or else the long-necked beast will open her door and come out and into mischief. The best type of bolt is illustrated above.

Panels of clear PVC can be incorporated into the roof or walls to allow for extra light, and any windows within reach of the goats should have the glass replaced by the same kind of clear plastic, which may save accidents. All partitions and gates should be at least 4 ft. (1.2 m) in height, otherwise youngsters could jump out at will.

Artificial lighting of some kind must also be provided, because you will only have sufficient daylight for your work night and morning for about half the year. If it is not possible to have mains electric light, battery lanterns are very good. If you have your own charger you may be able to use a car battery to power your lighting. Storm lanterns are useful because they can be hung up without fear of their being knocked over. Most naked lights present too great a fire risk with so much straw and hay about.

Your goat may kid at night, so you must be able to see clearly what is going on. You will need one light to every two or three pens, or else a movable one on a flex.

Pens

Life is made simpler for all by separating each milker into her own pen. Each animal gets her own share of the food without having to fight for it, and it is easier to get in and out for milking or to let the goats out for grazing. Goatlings and kids may be kept in twos and fours so long as they get on together and none is bullied. But do remember that two animals really means three pens if one animal is going to be in kid, because you will need a third pen to house the kids until their future is decided.

Milkers' pens need to be at least 6 ft. × 4 ft. (1.8 m × 1.2 m) so that a large goat can lie comfortably and stretch at kidding time.

Partitions between pens can be made from off-cuts from timber yards, affixed to framing of 2 in. × 2 in. (51 mm × 51 mm). Otherwise one of the harder types of wall board can be used, but not anything soft enough to be gnawed into holes. There are many types of suitable building materials such as concrete, breeze blocks, weld-mesh or similar netting fastened on to strong framework. Slats used should not be more than 3 in. (76 mm) apart, as small kids can get through anything larger than that, and if the occupant of the pen into which they scramble does not like them, they can be injured. Again, if the kids reach their dam, you will find your milk yield considerably reduced at the next milking. Whatever materials you decide on, make the section facing the door solid, to stop the wind blowing directly on to the animals.

Food Racks

It is possible, of course, to keep your goats collectively, in which case you must make a feeding heck to ensure that they all get their food. A feeding heck is a large board along the front of the pen, through which keyhole-shaped holes are cut, one for each animal. Outside is a ledge with holes to hold the food pails, and once the animals have their heads through the keyholes and are feeding, a bar is lowered the whole length of the pen, keeping the animals' heads at the pails.

Alternatively you can affix short

Feeding heck. The bar is raised when you are ready to allow the goats to come away from the pails

lengths of chain at intervals around your building, at such a height that the animals may stand or lie down. A collar and spring clip keeps each goat in her place where her food is given to her, thus preventing raiding by bullies: there is always one boss amongst a group who will take just what she wants unless she is prevented from doing so. The goats should be released after feeding.

Pails for water or food can be placed in a loop of strong wire which is attached to the wall or door by being hooked into two screwed eyelets. The loop will stand out straight when the pail is put in position and drop back flat against the wall when not in use. If a hole large enough for a goat to put its head through is cut in the door about 9 in. (228 mm) above floor level, and one of these rings placed outside the door, you can give both food and water without

21

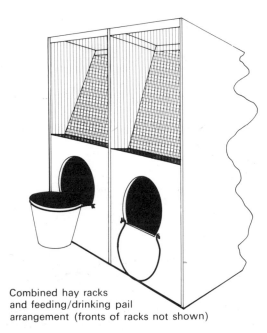

Combined hay racks
and feeding/drinking pail
arrangement (fronts of racks not shown)

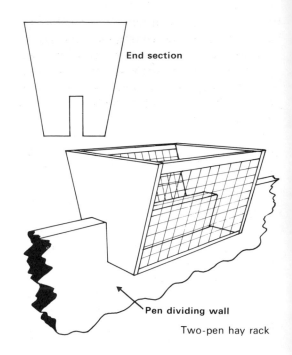

End section

Pen dividing wall

Two-pen hay rack

entering the pen. Alternatively, a length of light chain or cord can be fastened to the pail's handle and hung from a cup hook placed high up, so the goat cannot injure itself on the hook; this does prevent the animal tipping its pail over or carting it right to the back of the pen. You should remove empty pails, because goats like to play at mountains and will turn an empty bucket over and stand on its bottom, which in no way does good to the bucket.

Hay racks can be bought from farm suppliers or made by any handy person from woods with slats 2 in. (51 mm) apart. Alternatively they can be of wooden frames covered with mesh 2 in. (51 mm) square; but if the mesh is any larger the hay will be pulled through and wasted.

The most useful sort of rack is the half-rounded type with a flat back made of weldmesh, and a floor of plywood. This hangs flat against the wall or partition. Alternatively you can use a wedge-shaped rack with solid wooden ends into which a slot is cut. This fits over a partition, and thus supplies two pens. The sides, which can either be slatted or mesh covered, should be about 20 in. (508 mm) deep. The width at the top should be about the same, or slightly wider, and decrease to one-third of this size at the base.

Hay nets should not be used for goats. They are such active creatures that they are apt either to get a leg caught in the nets and dislocate a shoulder, or one may even get its head enmeshed. An animal can be strangled by the netting if no one is at hand to disentangle it.

Hay should be stored under cover as it absorbs water and will become fusty and damp quickly; it is now so expensive that it should be preserved in the best possible condition. If no building is available, cover it well with polythene fastened down on all sides. Never allow the hay to come into contact with the ground. Bundles of sticks, lengths of wood, even a deep layer of straw will help to keep an airway beneath and so prevent dampness and mould. Once damp, the hay will be wasted, because no goat will eat it.

Straw should also be kept well covered. A wet bed of straw is no use to any beast, and the associated dust and spores can also give the goatkeeper Farmer's Lung, a very nasty chest condition.

Although goats like a dry bed, they also like it to be thick. So do not clean out every week, this is wasteful of straw; simply straw the top to keep it dry and clean, and clean out monthly, or possibly at even longer intervals during the summer when the pen is not used so much during the day.

Outdoor Fencing

Once they get used to the idea, most breeds of goats will settle to whatever kind of restraint you decide to use. Anglo-Nubians, however, do not tether at all well, and become very vocal if they feel ill-used.

For many years I have used cattle fencing. This is 45 in. (1.1 m) high and has seven strands of wire, the bottom sections being closer together than those above. The vertical strands are about 12 in. (304 mm) apart. The posts should be about 10 ft. (3 m) apart and knocked 1 ft. 6 in. (457 mm) into the ground. The netting must be strained on to the posts and stapled in at least three places to keep it taut. The fencing must be strong: the goats will rub back and forth along any weak spot until it bulges and they can climb over.

Chain link is often quite satisfactory, though as it is extremely heavy it needs

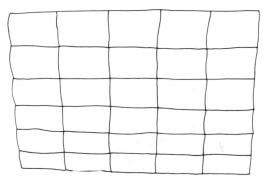

Cattle fencing

more and stronger posts. Its major disadvantage is that it costs more than most of us can afford. Electric fencing is very good once the goats have been taught to respect it. This can be done by introducing them in a fairly heartless manner. You can turn the current on, then without touching the wire yourself, you hang a wet twig or kale plant on it; when the goat goes to nibble the plant the current will give it quite a shock, as both its mouth and the twig are damp. It sounds hard, but it is no use being too sentimental about it,

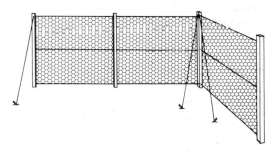

Section of chain-link fencing showing lines of wire to hold it firm and stay rods at each corner. Although other types of fencing can be stapled on to wooden posts, chain link is so heavy that it really needs posts of metal. Posts in all cases should be about 9 ft. (2.7 m) apart and driven well into the ground

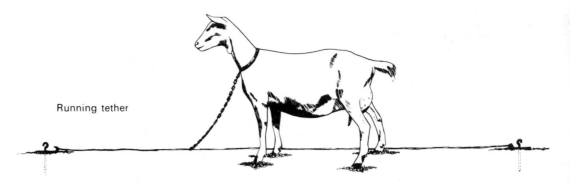

Running tether

otherwise you will spend many hours chasing an animal which is spoiling your garden, or worse, your neighbour's. Bad fences make bad neighbours very quickly. There should be three strands of live wire, one each at the knee 15 in. (381 mm), chest 18 in. (457 mm) and head height 25 in. (635 mm) of the goat. A fourth strand about 4 in. (101 mm) above the ground and not carrying any current will prevent small kids getting under the electric fence, but will not get 'shorted' by grass touching it. I should not care to rely on that kind of fencing for my goats. Flexinet is now available in a height suitable for goats and horses – 43 inches high, including 7 electrified strands, with a netting which increases in size of square according to height, so keeping in all goats from the smallest kids to the largest males. I use this, with excellent results, and have not had any animals escape from it. After coming up against it they very quickly decide it is much more comfortable not to do so. It is very effective and not expensive.

Tethering

Tethering is a last resort with all breeders and many will not sell stock to homes where they will be tethered. Goats hate the rain and may panic when it starts; if several are kept, it is no mean task putting them out and getting them in, plus moving them a couple of times daily to a new area of grazing. Always remember to keep the tethers away from bushes or large tussocks of grass, as they can soon become snarled and tangled. This can cause strangling, especially if a goat is being harried by a dog or teased by children.

If you must tether, running tethers are the best kind to use. They consist of two iron pins about 20 in. (508 mm) in length, with a flat disc of metal surrounding the head; into the edge of each disc is threaded a piece of strong wire, and on to that is slipped about 3 ft. (0.9 m) of chain with a swivel at both ends and a spring clip to go on to a collar round the goat's neck. The goat can graze a short area either side of the wire for a distance as long as the wire itself. Twice daily the pins should be moved sideways three times the length of the chain; it is necessary to prevent overlapping because goats will not eat from dung-fouled places.

The straight lines of this type of tether are far less wasteful of space than is the traditional manner of tethering with a central pin, so that the goats graze in rings. If several goats are kept they can graze alongside each other without getting their chains mixed, as so often happens with central pinning.

Feeding

Success is bulk shaped: the art of feeding goats is to get as much good quality bulk into them as possible, beginning with the best hay you can find. It is a very sad thing that so many people are under the impression that a goat can live on fresh air and paper bags; they could not be more wrong—goats are very particular about food being clean. Fusty damp hay or contaminated green, or in fact any kind of food which is not good, will be ignored or tipped over and trodden on. You can eat dropped food if you wish, but do not expect goats to do so—they will just refuse.

When introducing any new food, bulk or concentrate, do it very gradually, because goats are cautious, conservative creatures. Any new food will be tasted with small nibbles and nosings to make sure that it is suitable; then if they decide it is all right, the goats will eat it with relish. You can then slowly increase the amount fed.

Hay, Pea and Bean

Hay is a must. The best liked is clover, but it is very expensive and not too easily obtainable; lucerne, beloved of racehorse owners, is also excellent, but again, hard to come by. Meadow hay containing plenty of herbs as well as grass is, I think, the best bet for the average goatkeeper, and should be obtainable in most areas. Seed hay—that produced from sown lays—is of course good. However, it gives very little variety as it is usually confined to one or two types of grass; moreover, it is mostly ryegrass, not liked by goats, and so should only be bought where no other is to be had.

Baled **pea haulms**, from areas where peas are grown for drying, can be bought for a very reasonable price, and help out the more expensive hay; goats love pea haulm as it is coarse and crisp. **Bean straw** from field beans is also sometimes to be had and, though not so good as pea straw (dried haulms), makes a change of food which keeps up the bulk. A change is very useful when the animals are in during bad weather.

Greens, Roots and Fruit

Goats will eat the outer leaves of most vegetables, but the most useful greens and roots are cabbage, broccoli, sprout stalks, tops from carrots and celery, beet tops (which must be wilted before use), kale, mangolds, turnips, artichokes, kohl rabi and potatoes.

Green food should not be placed in small racks from which goats can pull it on to the floor: cabbage, sprouts, kale and so on should be hung in bundles. When sprouting is nearly over, broccoli should be pulled up and hung by a cord so that it can be eaten all the way to the root. Cabbage may also be fed from pails, the greens either being cut up small or halved and placed cut side up. **Artichokes** or any round food such as small **turnips** and **apples** should be cut in half. A goat has only one set of teeth and a hard pad at the top of the mouth, so it is possible for a round object to slip down its throat and choke it. Never feed any green food frozen; it can cause abortion if given to inkidders, and scours to others. Thaw it first. Males must never be allowed to have **beet pulp** or **mangolds**, as they are too sweet and cause stone in the urethra. This stops the passage of urine, to the animal's great pain and distress.

Beet pulp is a most useful adjunct to winter feeding, giving considerable bulk just at a time when outdoor food has gone. It may be fed dry to young female stock and is well liked, being sweetened with molasses. But milkers, and definitely all goats in kid, should have the pulp soaked, otherwise its greatly increased bulk can cause trouble by taking up too much space in an already full body cavity. Soaked well with very hot water, but squeezed dry and placed on top of the ordinary concentrates, or dried off with bran or grass meal, it makes a warm and filling evening meal in cold weather. Never soak beet pulp in a galvanised receptable, as it will dissolve the galvanising and poison your animals too. No goat likes sloppy food, so the pulp is squeezed dry, but they do like to drink the water in which it was soaked, so let them.

Mangolds are liked by most goats, but should not be fed until after Christmas when they become really ripe, and then only to females. Although some animals will scoop out halved mangolds, mine prefer them chopped up; even the youngest kids can cope with small pieces. The way to use **Jerusalem artichokes** is to cut the tops and feed them to the goats whilst the roots are left in the ground until required—they stand frost very well. **Potatoes** should not be fed raw, but steamed, and can be used mixed with bran. Although some goats are very fond of them, not all are, so try a small sample first. **Kohl rabi** can be given cooked, and also raw along with the leafy top.

If you live within reasonable distance of shops, perhaps you can persuade the greengrocer or supermarket manager that he would really be most pleased to for you to remove all his **vegetable trimmings** and any speck fruit which he cannot sell. Such things as celery, watercress and specked oranges make a nice change for a yarded goat. Apples —windfalls, peelings and cores from those being frozen or bottled—will be greatly appreciated; remember, however, not to feed more than 1 lb. (453 g) per animal each day, as an excess can cause scouring and a drop in the milk quantity. **Pears** are also well liked. Fruit tree and **bush trimmings** are first choice for any goat, but see that you do the pruning: goats will eat wholesale if allowed near the trees on their own, because, of course, they are naturally browsers, not grazers.

Concentrates

This is the name for the meals fed. These are various cereals, the most used being bran, crushed oats, flaked maize, some rolled barley, all of which are called 'straights' and are used mixed with the addition of proteins to make a balanced ration. The requirements of a milker are four of carbohydrate to one of protein; it is possible to mix your own, have the miller mix them for you, or buy the mixtures, or nuts, made specially for goats (see note on page 33). Since the goats are such bulk eaters, I

Analysis of Various Acceptable Foods

	Starch equivalent	Digestible protein	Dry matter	Mineral state	Remarks
Tree leaves					
Ash	14.1	1.6	20	high lime	laxative; much appreciated
Elm	11.1	0.5	20	medium lime	most liked; good for butter fats
Willow	7.6	1.9	20	high lime	well liked
Vegetables					
Artichoke	16.2	2	32	medium lime	well liked
Cabbage	6.6	1.8	11	sufficient	acceptable
Chicory	5	0.7	11.5	all required by goats	acceptable
Kale, thousand headed	10.3	1.7	15	sufficient	greatly liked
Kale, marrow stem	9.1	1.7	11	medium lime	greatly liked
Roots					
Artichoke	16.4	1	20.4	low	well liked, feed chopped
Carrot	8.8	0.8	13	low	accepted chopped
Mangold	6.2	0.7	12	low	liked, chopped, not suitable for males
Swede	7.3	1.1	11.5	low	may be fed to males
Cereal grains					
Barley	71.4	7.6	85	high phos.	acceptable for growing stock
Oat	59.5	8	87.7	high phos.	well liked
Maize, flaked	77.6	7.9	87	high phos.	well liked
Wheat, as bran	42	10.9	87	high phos.	very acceptable
Oil cakes					
Groundnut	77.5	42	89.7	high phos.	acceptable
Linseed	74	25	88.8	high phos.	usually liked
Soya bean	68.9	38.8	85.5	high phos.	acceptable
Bulk food					
Beet pulp	60.6	5.3	90	high lime	well liked and a good filler

would always prefer the mixtures, and if you choose to have a ready mixed ration, I would suggest buying 25 kg bags and the same weight of bran, and mixing the two. Bran is itself correctly balanced, so can be added in any quantity without upsetting the balance. The concentrates should be divided into two parts and fed morning and evening before milking; if allowed time to eat her food—about twenty minutes—the animal will then stand and cud contentedly whilst being milked.

Very few goats eat the nuts with pleasure; a few of them do so, but they soon get tired of them. To save wasting them, they should be added to a mixture of crushed oats and bran and flaked maize, in about equal portions by weight. 1 lb. (453 g) of concentrates is needed daily for maintenance, that is, to keep the body replacements correct. In addition, 3 lb. (1.6 kg) are needed for each gallon (4.5 l) of milk produced. Remember that a goat carrying two or three kids is working just as hard as a goat giving a gallon of milk; do not cut her food down. The yield will drop towards the autumn to about two-thirds to half of the summer quantity, but providing she is well fed, the animal will continue to give this throughout the winter. Her output will begin to increase with the coming of the lighter days, especially when the new grass appears. A most useful attribute of the goat is this 'running through', an ability to milk for two seasons without having to kid in between.

Sample Rations

The following chart gives some suggested sample rations. The cereals, being approximately equal in feeding values, are interchangeable; all parts are by weight.

		Part
1.	Linseed cake	1
	Kibbled beans	1
	Flaked maize	1
	Crushed oats	1
	Bran	1
2.	Crushed oats	2
	Kibbled beans	1
	Maize gluten feed	1
3.	Decorticated nut cake	1
	Malt kulms	1
	Flaked maize	1
	Crushed oats	3
4.	Crushed oats	3
	Broad bran	2
	Split peas	2
	Locust beans	1
5.	Middlings coarse	2
	Crushed oats	2
	Bran	2
	Molassine meal	$\frac{1}{2}$
	Kibbled beans	2

The rations are based on 1 lb. (453 g) for maintenance and a production quantity based on the animal's output, allowing $3\frac{1}{2}$ lb. (1.6 kg) for each gallon (4.5 l) produced.

Newly Kidded

Remember, however, that a newly kidded animal may have straights, i.e. one or other cereal, or mixed, but no protein, i.e. cake, peas, beans or soya meal, for a week after kidding; bran is excellent, and should always be included at that time. Too much protein can push up the milk production too fast, and cause milk fever. By the end of the first week the animal can have some of her normal feed added; gradually it is increased and the straight cereals decreased, so that she receives her normal food by the end of the second week. Her output will usually rise steadily for about six weeks, at which time most goats reach peak production; some, of course, take more time, whilst others reach it in less.

Young Goats

Younger animals get some of the same foods, but proportionately less. Kids begin to nibble meal from two to three weeks of age, and should be given just a small handful in a bowl, as some of it will assuredly be wasted. Shortly, however, they will grow to like the food, and in that way get used to variety, so that when adult they will eat whatever is offered to them.

Encourage the eating of bulk foods by gradually increasing their amounts. Hay will be nibbled from a few days, and soon will be taken in fair quantity; green food in small quantity should be given, but never directly before a milk feed. To prevent bloat or scouring, at least half an hour should elapse from when the kid eats grass, etc., to when it is given a bottle.

Goats are called kids until one year old, whether they are males or females. At that age the females become goatlings, the mischievous and inquisitive delinquents of the goat world. As the kids grow, you should increase their meal ration and decrease the milk feeds, so that by six months they are having about 8 oz. (227 g) morning and afternoon. Keep them at about that amount until they are a year old, with more bulk feeding. After a year, make a further gradual increase up to about 1 lb. (453 g), which should be reached when they are ready for service in the

autumn. Once served, the increase goes on, so that by a fortnight prior to kidding, the goatling is receiving a full milker's ration to keep her growing well and supply the needs of the kids she is carrying. During the last two weeks, at least half the food should be bran. Steaming up, as for cattle, is not a good thing for goats.

If you mix your own foods, the added protein will need to be one of the following: kibbled beans, crushed peas, decorticated ground cake (expellers), linseed cake, cotton seed cake or soya bean meal. The first two, being grown in this country, are more likely to be obtainable than imported cakes, and, one always hopes, more cheaply.

Growing Food

Many folk are now returning to the idea of helping themselves in many ways, and anyone reading a book on keeping goats is probably in their ranks. No doubt you will wish to use a portion of any land you have at your disposal for growing food, especially that which can be used by humans or animals in the family circle. One thing that you will find to be a great boon is the manure from the pens; it really makes things grow in a truly astonishing manner.

Should your garden, or whatever area you consider using for your goat project, be less than $\frac{1}{4}$ acre or $\frac{1}{8}$ hectare, do not use it as grazing, it is far too precious. Make an exercise yard for your household providers and feed them their green food in racks in the yard, or house, as weather dictates. This yard does not have to be large; it will be satisfactory if it is, say, three or four times the area of their shed.

If left on a small grass area, the goats will become very 'wormy' and the ground will become goat sick. Do not attempt to make use of any top growth in their yard: in fact a concreted yard would be much better. You are just aiming to ensure that the animals are not always confined to their stalls, but can walk about as they wish.

The remainder of your land will supply a great deal more food if it is utilised for crops rather than for grass. In many cases the crops are dual purpose, in that the heart and better parts are used in the house, whilst the outer leaves, cores and stems will be happily consumed by the goats.

The most useful crop is **kale**, grown for winter use. The marrow stem kale is best up to Christmas, after which winter frosts split the stems and rot them; this should be followed by thousand-headed kale, which will stand the rigours of weather better. It is usually thought that if you allow one kale plant per animal per day from October to April, plus any other green food or roots, you will give good overall cover for your animals.

You will find it beneficial to include a few rows of **lucerne**, which has to be sown in clean land and kept clean until it is going well. Once established, it stands without replacement for up to seven years; and after the first season, when it is cut once, you will get at least three cuts each year, because it grows so fast that it appears almost to come up behind you as you cut a part row each day. You should cut lucerne before it flowers and then leave it to wilt over-night before feeding it to your goats.

Agricultural **chicory** is also a good long-standing crop. It grows fast and tall, can be cut up to five times a year, and is also said to be the only plant which supplies all the minerals which a goat requires. As its roots go down a prodigious depth for water, it is pretty drought-resistant and an excellent crop for sandy areas.

Comfrey and **artichoke** are good to feed on wet days, because as their tops are coarse and hairy the rain does not hang on the leaf, as on softer herbage.

A new plant has recently become available. Called **tyfon**, this is a cross between a stubble turnip and Chinese leaves. It is a very useful leafy plant which, if sown in June, produces a cut in 10 weeks and a further cut in October. If left in situ, providing the winter is not too severe, it will provide a third cut in late February or March.

Most goat clubs obtain seeds of such crops as kale, fodder beet, lucerne and mangold, and supply their members with the one or two ounces. This is a most useful service, because these are farm supplies and are usually only to be had in far larger quantities than are needed by the average goatkeeper.

Making Hay

If you have grazing as well as arable land, and are fortunate enough to be able to reserve part of it for hay, a very good way to make the hay is to use tripods. These are constructed from three heavy pieces of wood, placed tentwise, with their top ends lashed firmly together. The uprights are secured by the addition of cross pieces, tied to the uprights and crossed at the corners; at least two of these side cross bars should be added to each side.

When your herbage is cut it should be wilted for several hours, then laid across the side bars to hang; you can add more as you cut it, but you must make sure

Hay tripods

The tripod uprights can be any height and may be made of any wood you happen to have—even branches from which your goats have stripped the leaves and bark. An equal distance between the three legs makes for balance. The side pieces can be of light wood, wire or even plaited twine; they are just to support the herbage.

On the ground between each set of legs place an object, such as a pail or a log, and when the tripod is finally loaded, remove this object to allow air to go up the hollow centre, preventing moulds forming. The hay must be very well dried, i.e. cured, then removed and stored in a covered place. Tripods may be filled several times during the summer, with hedgerow and ditch weeds as well as grass and nettles; and of course any number of tripods for which wood is available may be used. It is quite the finest hay you will ever make

that a gap is left up through the centre. This makes very sweet hay. You may, of course, add herbs (or weeds, whichever you prefer to call them) from the hedgerow to make for variety.

If you have no sturdy poles, try and stretch a piece of netting across some sticks just clear of the ground. Small amounts of herbage laid across the net-

ting will get a good current of air all round. When the hay is dry, pack it into sacks and store away from damp.

Wild Plants

If you live in the country, you will be able to supplement the diet of your goats with lots of food free for the collecting.

Among those suitable are:

agrimony	meadow sweet
burnet	plantain
campion	sheep's parsley
clovers	sorrel
coltsfoot	sow thistles
cow's mumble	tansy
dandelion	vetches
docks	watercress
grasses	wild chervil
hogweed	wild parsnip
hops	yarrow

Ivy makes a good tonic, although the berries should not be fed. First and foremost among the useful plants, is, of course, grass. When first grown it is very high in protein; later in the summer, the protein content diminishes, and in July and August you will need to increase the intake of protein to balance the lack of it in the grazing. Branches from ash, elm, willow, hazel, dogwood, elder and several other trees are relished as a real treat. However, although all of these can either be fed from the plant or cut, hawthorn, briar-rose and brambles, well loved though they are, should only be allowed when they can be eaten from the plant. The branches of these bushes can cause a great deal of trouble if their thorns pierce the goats' udders or get between the two claw sections of the hoof. Blackthorn will also cause grazes which can turn to festering sores.

Some of the edible wild plants you can find for your goats: (1) tansy; (2) salad burnet; (3) watercress; (4) dandelion; (5) hogweed; (6) coltsfoot; (7) yarrow; (8) hops; (9) corn sowthistle; (10) curled dock

Poisonous Plants

Some plants are poisonous and should never be fed to goats, among them the garden rhubarb. Laburnum is a tree which is deadly in all parts—leaf, seed and bark. Evergreens such as fir, spruce, etc., are all suspect because they cause scouring; privet berries are bad, although some green is not too great a hazard. A list of poisonous plants would include the following:

alder	laburnum
box	rhododendron
bracken	rhubarb
broom	rushes

bryony (black and white)
daffodil
dog's mercury
foxglove
ground elder
hemlock
nightshades:
  black
  deadly
  woody

spindle
thornapple
water dropwort
wild arum
wild clematis
yew

The Ministry of Agriculture and Fisheries produces a very useful book, Bulletin No. 161, 'British Poisonous Plants', which gives you the plants, their effects and antidotes, if any; all goatkeepers should have a copy.

Note

Many well-known milling firms now make concentrate mixtures specially for goats in 25 kg bags, thus enabling customers to buy reasonably small quantities, as long storage is not recommended, especially in summer.

Breeding

The recognition of the oestrus or 'heat' period is often difficult for those starting in goatkeeping. Most books say that you will be able to tell because 'she will do' some particular thing. In fact the books should really say that she may show any—and in all probability she will show at least two—of the following signs:

- a restlessness and disinclination to eat
- a continual wagging of the tail
- a sticky tail
- a slightly pink and puffy vulva
- a particular calling (from which comes the expression, 'she is calling', or ready for service); this is a continual bleating from a usually quiet beast.

Often you hear the rhythmic tapping of the tail on wall or partition before you see the animal at all. It is not the quick swish, as if to remove a fly, or the rapid 'pleased to see you' way. Combined with any of the above factors, in September or October, this tapping is almost certainly the sign of heat. Should you wish to mate the animal, and are not quite sure of your findings, mark the date and watch her in twenty-one days, the number of days between the heats. If unmated, the animal will call at the three-week interval from September to February, and the time she takes over it will vary with the season. In September, it will be just a few hours, but in October and November it will last for forty-eight hours. If you have decided to have the goat served, do go as soon as the heat is detected. If you wait until the second day she may be 'off' and you or the stud goat's owner will have a useless journey.

All female goats, from the youngest kid to the oldest milker, will come into season, but it is not usual to mate the youngsters until their second autumn, when they are about eighteen months old, and will kid around their second birthday.

Finding a good male can be a problem, but the British Goat Society produces yearly a Stud Goat List which includes males of all breeds and covers all parts of the British Isles. It also states the fees required, and whether the stud's owner will board goats for service or transport the male to you.

The local goat club will no doubt also have its own list. So you could get to know the whereabouts of the nearest suitable male by either joining a club or asking the secretary for information.

Fees for stud vary with the age and breeding of the male from about £2.50 to £10 (1977 prices), but you get a very good choice of really excellent males.

If you are going to take your goat to the chosen male you should try and make an appointment rather than just arrive with her, because it may be quite impossible for the male's owner to be on tap all day. Again, the male may just have been used; if he is young, he will probably not be allowed to serve a second female within 24 hours—though most are only too willing. The service is not a prolonged affair, it is really all over in seconds. Most stud owners will allow a second service to be given before you and your goat leave for home, especially if you live some distance away.

Once served, the goat will resume her normal way of life. Watch her at twenty-one and forty-two days to be sure she has not 'turned', i.e. come into heat again. If she does turn at either of these times, you can take her back to the male and get a free mating. However, it is seldom necessary to go back, because these males enjoy their work. After the second period you cannot get a free service because the failure will not be the fault of the male; there will be something slightly wrong with the female.

The gestation period is 152 days, but the animal may successfully produce her kids any time after 145 days.

If she is already a milker, her yield may show a slight increase for a time after service, but by eight to ten weeks this will begin to drop, and should be encouraged to do so. Milk irregularly, leaving some milk in the udder at each milking; then when the yield has dropped still further, to say 2 lb. (0.9 kg) daily, milk only once a day, then every other day, and so on. Try to have the goat dry for eight weeks prior to kidding to give her time to build up reserves for the next lactation. Some heavy milkers can never be got to dry off: these should be well fed and just eased, not stripped out.

Goatlings will show no apparent changes for about ten weeks after mating, then you will notice that they are getting more solid amidships; increase their ration gradually from then on. At about six weeks before kidding, a goatling's udder will begin really to show what it is going to be like in shape and, later on, in size. Some goatlings get very tight udders, and if one gets hard, pink or shiny, remove enough milk to make it soft and pliable, but do not strip out.

The signs for both first and subsequent kiddings are the same.

One more thing; many people are worried when the goat develops a blood discharge about a week after kidding, though until then she has had none. Do not worry, it is all quite as it should be. This discharge is usually quite slight but continuous for up to three weeks, when it will disappear of its own accord.

That is the purely physical side of breeding, but there is more to it than that, because your aim should always be to improve your stock. You should study your goat yourself, and also get some knowledgeable person to look her over for you and indicate her weak points, as one seldom sees all of them for oneself. Having found the deficiencies, locate a male from a family of goats which is good in the very areas where your own animal is poor; mating these will ensure that there is at least a chance that the doe kids will be an improvement on Mum.

Among the faults which should be cleared up are 'cow hocks', the condition in which the hocks are too close together and so tend to cause bruising of the udder as the goat walks. Another common fault is too great a slope to tail, which usually means there is insufficient room for a wide, well-attached udder; so a nice level top line should be sought. Length of body and many other of the goat's physical qualities (listed on page 15) can be inherited from the male. So really study the list of stud goats and have a good look at female relations of these males. You will soon see whether he is from the kind of herd which you aspire to own eventually.

You can grade up from an unregistered goat to the *Herd Book* by taking successive generations to good registered males. It is best to decide what breed you like and aim for it, even if your starting position is far down the line. 'Strive for the highest' is a good motto.

The artificial insemination scheme is now well established, and "straws" of semen are available of most breeds from COBS – Caprine and Ovine Breeding Services. You can obtain the name and address of your nearest inseminator, together with information on costs for straws, the actual insemination, and travelling expenses of the inseminator, from Pat Young, Secretary, COBS, Sykes Halt, Myerscough Road, Balderstone, Blackburn, Lancs, BB2 7LB or from Mrs A. May, Priestland Farm, Claygate, Tonbridge, Kent, or Mrs P. Carter, Metway, Southwold Road, Brampton, Beccles, Suffolk.

Kidding

A fortnight or so before kidding, clean out and disinfect the pen in which the goat will kid. Give her plenty of bedding and prepare a deep box—a tea chest or similar—in which to place the kids. The following items should be collected and placed in a large polythene bag and kept near where they will be needed:

- a couple of old towels, or rolls of kitchen paper
- iodine
- a nail brush
- a piece of soap
- an empty wine bottle
- a lamb teat
- a clean pail
- Dettol or similar disinfectant.

The signs of imminent kidding are roughly these:

- a great softening of the area at the base of each side of the tail, on the back of the goat
- the tail itself rising up
- the flanks hollow as the kids drop into position for their emergence, the udder filling up rapidly (it may have filled up considerably prior to this), but when it becomes really large and rather harder than previously, you will know that there are just a few hours to go.

A goatling will be making her first udder, and this will gradually grow over some six weeks, whereas the udder of an animal which has kidded previously will not begin to fill so early.

Restlessness, standing up, lying down, digging up the bedding, talking to her sides and grinding her teeth—these are all signs that you should keep an eye on your beast. It may take several hours, but finally she will begin to strain, and ever stronger contractions will drive the kid towards its final exit. If all is well you will see two hooves, one slightly before the other, with the tip of the nose lying atop.

Though it takes a little time, the mother should be able to cope perfectly well. No doubt there will be a loud wail as she pushes the kid's head through the bony outlet, but once this is achieved, the remainder of the kid will follow rapidly. If the mother is standing, help to lower the kid to the straw. Place the kid close to the mother's head, so she can lick it dry.

As there is usually more than one kid, be prepared to move the firstborn, as the mother can easily lie or tread on the first kid whilst delivering the second or third. Pick it up in the towel, give it a good rub and then place it on a nice bed of straw in the large box you had ready.

Each kid should be taken in its turn and kept warm until the mother has dropped the placenta (afterbirth), which usually takes about two hours; she will stand beside the box, which can be in her pen, and talk to and nose the kids. Put some iodine on the cord under the kid's stomach to prevent germs invading the raw area and causing navel ill,

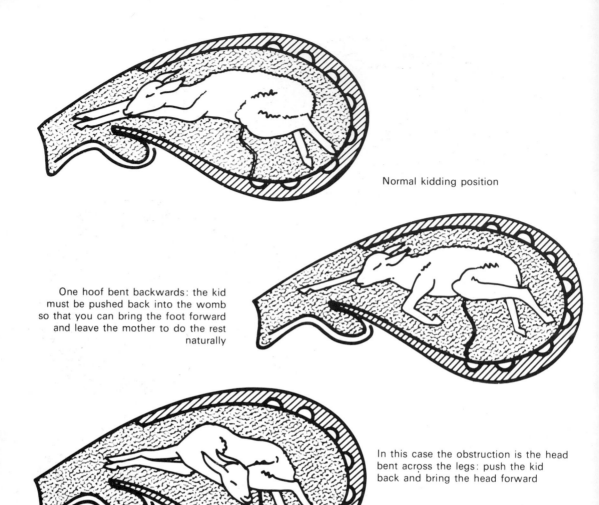

Normal kidding position

One hoof bent backwards: the kid must be pushed back into the womb so that you can bring the foot forward and leave the mother to do the rest naturally

In this case the obstruction is the head bent across the legs: push the kid back and bring the head forward

A straightforward breach presentation: as the rear legs are extended, the goat will manage naturally. When the rib cage emerges you should place your hands around the kid's chest and prevent it gulping liquid

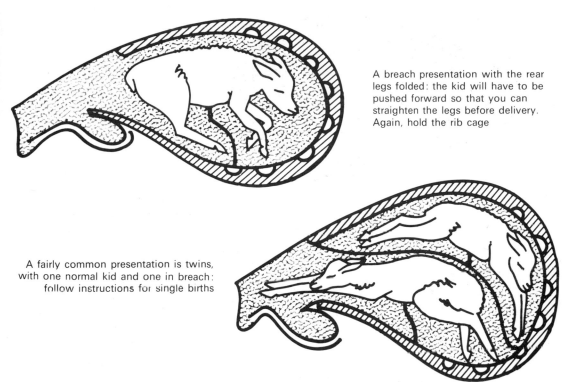

A breach presentation with the rear legs folded: the kid will have to be pushed forward so that you can straighten the legs before delivery. Again, hold the rib cage

A fairly common presentation is twins, with one normal kid and one in breach: follow instructions for single births

etc. Wipe the goat's udder with a warm, damp cloth and squeeze some of the thick yellow colostrum into the wine bottle, which should previously have been warmed with water. Place a lamb teat into the bottle top and give each kid a feed. This both ensures that they get the various antibodies, minerals and laxatives contained in colostrum, and also warms and comforts them.

Once the placenta has dropped, remove it to prevent the goat getting indigestion as a result of eating it to clean up. Never pull the hanging cords, etc., as this will cause haemorrhage.

Now clear up and spread dry straw. Turn the kids' box on to its side kennel-wise; the kids will use it to sleep in and to keep out of draughts and from under Mother's feet. When removed to another pen along with the kids, the box gives some continuity to their life.

Abnormal Kiddings

Having read about a normal kidding, I expect you would like to have an idea of what could go wrong and what you should do about it.

Should your goat go through all the early stages of kidding, and then strain hard with no obvious signs of anything happening, wait ten minutes, and if there is still no sign of hooves or nose, get assistance. An experienced goat-keeper or shepherd would be fine, a vet obviously splendid but costly. You may have to solve the difficulties yourself, as follows.

First wash the goat's hindquarters with warm disinfected water, then roll up your sleeves and use soap and disinfectant to scrub your hands and arms right up to the elbow. Soap one hand well, make your fingers into a funnel shape, and slide the hand right into the goat's vagina, where you will feel the kid blocking the way. Push it gently but firmly back into the womb. When you have room, feel for the offending obstruction and remedy it. The most likely trouble is a hoof turned backwards, in which case you must bring the foot forwards with the other so that the goat can bring the kid out naturally. Another fairly common problem is a kid's head being turned back over its shoulder; in this case you bring it round on top of the front feet and gently guide it to the opening so that it can come forward.

In twins or triplets, one kid is frequently a breech, that is, coming backwards, either with the legs stretched out or else tucked up under the body. In the former—the extended—case, the goat can usually manage. However, as the rib cage emerges you should place your hands around the chest to prevent the kid trying to breath as the cold air hits it; otherwise it could suck in liquid and drown before the birth is completed. In the second type of breech, the kid should be pushed back until the legs can be brought out one at a time towards the orifice. Then, holding the hooves above the foot, pull gently towards the floor, but only when the goat strains.

Once born, a breech kid should be held head down to drain for a few seconds. Its nose and mouth should be wiped, then it should be given to its mother to be dried and cleaned. An apparently dead kid, whose heart is beating but not breathing, can be brought to by administering 4.5 cc of Respirot into its mouth very quickly after birth. This causes rapid expansion of the lungs whereby a breath has to be taken. The kid should speedily recover. Following an assisted kidding, I would always have a goat given an injection of antibiotics; however careful you are, it is easy for infection to be carried in and for metritis to develop. A veterinary visit really is a great help, because frequently the placenta does not come away quickly. Although a goat's placenta should never be pulled away (it causes haemorrhage), a goat's uterus closes very quickly, so she must clean within twelve hours or risk a fatal septicaemia.

A sympathetic person to hold the goat steady for you is a great help. Do talk to the animal; a goat likes and trusts humans and really will respond when you tell her what a good girl she is and how much better she will be very soon. Goats are odd creatures, and quite expect to die whenever they have a pain —so you simply must raise their morale at all costs. Take heart, however: it is seldom that help is actually required.

Kid Rearing

First, make sure whether you really want to rear the kids. Unless you are going to use male kids for meat you will only keep females. Consider your needs for milk and decide whether you can spare the four pints which are required for each kid daily.

Sexing Kids

Kids are very easy to sex, unlike kittens, which I think are most difficult.

Firstly, if inspected below the tail, a female kid will be found to have two openings, the anus, from which dung will be passed, and the vagina or vulva, from which it will urinate and at a later date produce its kids. The lower end of the vulva has a lip, the clitoris, and if this lip is just neat and small all is well;

but an intersexed kid will have small swelling about the size of a pea or smaller. Such a kid will never be any good, but as intersexing is a rare condition get a knowledgeable person to check the kid for you.

Male kids have only one opening below the tail—the anus. Under its body is the penis through which it will pass water, and between the back legs is the scrotum, a small fold of skin which will contain the testicles. All kids of both sexes have two teats just in front of the back legs on the body. Inspect carefully to see that there are not more than two, as occasionally a supernumerary teat is to be found. This is a hereditary fault, and although some people will say that one teat can be removed, the affected goat should be culled. Otherwise, any offspring will probably show the same abnormality, and there is also a danger to the goat itself because there is sometimes a small hole left where a teat is removed: this can be a source for infections such as mastitis.

The Milk

The goat's milk is not of normal constituents during the first four days from kidding, and is only suitable for the young for which it is intended. There is no point in removing the kids from their dam until the fourth day, but they may well not need so much milk as is there. However many kids are born in a litter, they usually all feed from one side. You must, therefore, keep an eye on the unsuckled side, making sure it does not get 'overstocked'—too tight. Milk off some so as to keep both sides even and soft, but *do not milk out* until the fourth evening, or the goat may well get milk fever from too great a reduction in her system's calcium.

Once kidding is over, feed the mother a warm bran mash, a drink of warm water, plenty of hay and some greens. Apart from making sure the kids are suckling, there is nothing else to do except keep an eye on your family.

Feeding Kids

More often than not, kids are removed from their dam on the fourth day and fed by bottle. This makes sure that you know how much the milker is giving and how much the kids are receiving; it also makes for very tame and friendly little animals. Following their removal from Mum you will be able to sell your kids to anyone able and willing to feed them four times daily with milk. Milk substitute such as is fed to lambs or calves is also quite satisfactory, so long as the owner keeps to the makers' instructions for mixing.

In 1982 a new product called Kidolac was brought out. It is a Volac milk powder containing correct fats, vitamins and minerals and is especially good for kids.

Some people teach kids to drink from a bowl or pail, which has the advantage of quickness. On the other hand, because the kids guzzle, it makes for digestive upsets and pot bellies. On balance it is not really to be recommended.

Rearing on the dam is, of course, natural. However, many animals dislike getting no peace because they have their offspring at their heel all day and night. You will also find that you will have to strip out the dam very carefully twice a day when her kids begin to take less, otherwise the yield will go down to zero. Owners who are out at work during the day might reasonably leave the kids with their mother during

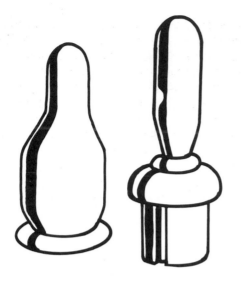

Bottle teats: on the left, pull-on type; on the right, plug-in type (note groove in plug to allow air entry)

these hours. They can strip out the milker in the evening and remove the kids overnight to a separate pen; in the morning they can take the milk for the house, before giving the kids back to their mother.

At about ten days, a small feed of hay should be given in a rack into which the kids cannot get their heads.

Young kids being bottle fed

(Remember that hay must never be fed in a hay net in which the kid can strangle itself; a box can be used if you have no rack.) About a week later, offer small feeds of meal such as the kids' elders and betters are having. In this way they will begin to eat and enjoy many foods, though they should not yet be fed much green food because, with the milk intake, this can cause scouring and bloat.

Once feeding well, kids go on growing strongly. Whether they are being kept for meat or stud, male kids must be removed from the female kids at ten weeks, because from then on they can give effective service and will certainly worry the female kids with their attentions. For the first few days milk will be taken at about $\frac{1}{2}$ pint (0.3 l) four times daily. At about three weeks to a month this will have increased to about a pint (0.6 l) taken four times daily. Then make the quantity $1\frac{1}{4}$ pints (0.7 l) in each bottle, given three times daily and keep up these three feeds, plus an increasing amount of meal, for five months. At six months reduce to one bottle night and morning, then after one month reduce to one bottle in the evening, for at least one more month, preferably until eight months.

Plenty of hay, bulk food and exercise will now be needed.

All this is for those who do rear their kids, but please make sure before embarking on this long and expensive experiment that the kid is worth it. Look well at it; make sure in the case of a female kid that it has no extra teats or swellings at the base of the vulva—this would indicate an intersex, which under no circumstances should be reared. Be hard, only keep the best. The total cost of the milk alone is alarming; and now think, that is why breeders want a good price for their kids. They have borne the expense of the rearing and have done the hard work.

Horns

Horns must be removed by the end of the first week of life, not at three months as in the case of calves.

It is now illegal for any unauthorised person to disbud kids, which, in effect, means taking your kids to the vet. When you do this, ask for a general anaesthetic rather than a local one, which can cause considerable upset, even meningitis.

Horns left on to grow can become dangerous. This is not because the animals are aggressive, but simply because if one brings its head up as you bend over, it can give you a black eye or bruised chin. Bumps between goats can cause injury and mastitis.

Milking

Milking is not at all hard, but an acquired knack is necessary. Where the goat stands—either on a milking stand or on the ground—should previously have been swept or swilled clean. The animal should be placed against a solid partition or wall, preferably with her head into a corner so that she cannot walk forward. The milking pail should be set on the floor, just in front of the udder. The pail should be made of metal, and although enamel or plastic will do, the fats will stick to the plastic somewhat.

Wipe the goat's udder and teats free of dust and dirt, using a J-cloth wrung out in warm water. With one hand to each teat, close each thumb across the palm of the hand to cut off the flow from udder to teat. Squeeze firmly downwards on one teat, *following one finger with the next* until all the teat is empty. *Never pull the teats* or draw your finger and thumb down from top to bottom as is done with cows, because this injures the very tender structure of the udder and can cause mastitis. Release the first teat and repeat the process on the other one. Working alternately like this you will get a jet of milk from each teat in turn. Do this as quickly as you can, otherwise the animal may get a bit restless.

When you have apparently removed all the milk, rub your hands from the top right under the body down towards the teat, and you will find more milk to take. Do this several times—this is called 'stripping'. Make sure that you strip all the milk, otherwise the goat will think that you (her kid substitute) do not want it all and reduce her yield. Again, the last milk contains the butter fats—so no stripping, no cream. Doing everything in the same order each time will cause the 'let down', when the most milk will be obtained.

 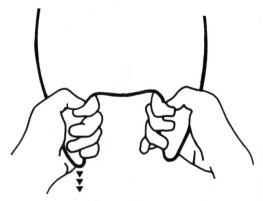

Milking. On the right-hand teat the fingers are one at a time squeezing the milk downwards and into the pail, while on the left-hand teat the thumb is in position to cut off supply from the udder above. Each teat is squeezed alternately and as fast as possible, speed coming with practice

Uses of Milk

Naturally you will wish to get the best from your investment, so firstly, feed your goats well. The second essential is to keep your utensils clean and sterilised by using a steralent, which can be bought from dairy suppliers. The same suppliers will also sell you filter pads.

Milk must be cooled immediately after being milked and strained, as bacteria will be rampant in a warm medium. A filter such as a tissue or paper kitchen towel between two nylon kitchen strainers is quite effective and cheap. The milk should be strained into the vessel with a well-fitting lid, and is then cooled by placing the closed container in a sink or bowl and under a tap with cold water running slowly. When the milk is cooled, leave the vessel in the cold water, or else divide the milk into bottles and place them in the fridge or cold pantry. (See note on page 49.)

Now that you have the milk you will discover that you can use it to make anything which you can make with cows' milk (only better). If you have a separator you will be able to make butter, if not you can prepare clotted cream. If you are going to do much dairying, you will find that a dairy thermometer is well worth lashing out on.

Clotted Cream

Place some milk in a wide-topped bowl, not a plastic one, and leave it to stand for twelve hours, after which you should warm it very gently until it crinkles, and then remove it from the heat and leave for twelve more hours. Skim off the cream: this can be used with fruit or kept to make butter. Set up the next bowl of milk and continue the process until you have sufficient to make your butter.

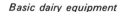

Basic dairy equipment
From left to right: one-gallon milk churn; milk carton for retail sale of milk; milk pail; two-gallon milk churn with strainer in position; churn lid

45

Butter

Take the cream, which has been obtained either from a separator or as clotted cream, cool well and beat it with either a wooden spoon or an egg whisk, or—least trouble of all—use an electric mixer. In a short time small grains will appear. Place muslin in a strainer or colander, tip in the butter grains and use very cold water to rinse them free of milk. When the water runs clear, take the corners of the muslin and gently squeeze out as much water as you can.

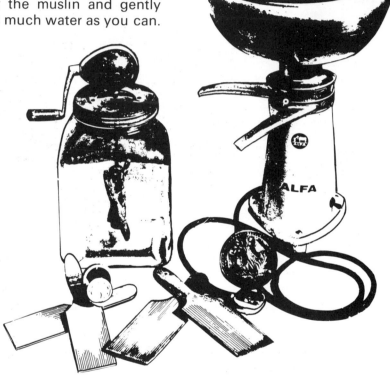

Butter equipment
A glass butter churn and electric separator behind a butter mould, butter print and butter pats (scotch hands)

Place the pat of butter on a clean table or board and beat it. Preferably you should use butter pats, but failing these, you can make do with two wooden spoons or spatulas, or even your hand. Sprinkle the butter with salt and beat until all the water has stopped weeping out. Then shape your butter and place it on a plate to cool and set. It will then be ready.

Goats' butter will be quite white unless you add some butter colouring, which is named Annatto and obtainable from your dairy supplier. There are two kinds of Annatto, one for butter and one for cheese; the butter variety colours the fats, and the cheese only the non-fat solids, so you must be quite sure you get the correct sort.

Cheese

The dairy supplier will also stock Rennet, an essential for simple cheese-making. Do not try to make sour milk cheese or you will be very disappointed. Goats' milk takes a long time to go 'off', and even then it does not sour like cows' milk, but simply goes bad. If you ladle your cheese curds into it. Tie the corners of the muslin together, and hang it up to drip for twelve hours in some place where there is a good current of air. After that, take down your cheese, undo the cloth and scrape sides to middle. Beat in some salt and, if liked, some chopped herbs such as chives. Your delicious cheese is now ready to use.

Cheese-making equipment
From left to right: Cheese press for six cheeses at once; ratchet press for small-holder type cheese; Pont l'evecque mould of wood with a dairy thermometer; bowl for setting milk, rack with draining basket lined with cheese cloth held by spring pegs; curd or cream skimmer

leave it till then you will have a disgustingly flavoured mess on your hands.

Simply place what milk you can spare into a flat bowl which must not be plastic, because it needs to be heated. Warm the milk to 90°F (32°C) and stand it in a place where it can remain undisturbed for a couple of hours. Stir the warm milk and add to it ½ teaspoonful of Rennet in 2 teaspoonfuls of cold water. Cover the liquid with a clean cloth, arranged so it will not dip in, and leave to set.

After two hours, line a colander or strainer with well-scalded muslin and

When you are ready to try more complicated recipes you should refer to the British Goat Society's excellent publication, *Dairy Work for Goatkeepers*. Costing only a few pence, this contains many recipes for cheese, butter, cream and yoghurt making.

Yoghurt

Yoghurt can be made easily without any special apparatus except a thermos flask. Slowly heat sufficient milk to fill the flask to 170°F (77°C). Cool it rapidly to 120°F (49°C) by standing the vessel in cold water. Place 2 tablespoonfuls of natural yoghurt in the thermos flask, fill it up with the warm milk, close and shake. Leave the flask overnight, then in the morning pour the liquid carefully into a bowl and cool rapidly. If you also use the flask for tea and coffee and it has a stopper of cork rather than plastic, cover the cork with a piece of polythene, otherwise your next lot of drinks will have yoghurt in them; these little bacteria creep into the cracks and are astonishingly persevering. To make a second lot of yoghurt, just add some more warmed milk to a little of your first batch. Natural yoghurt culture can be obtained from dairy suppliers, or else a carton of natural yoghurt may be bought from a grocer's shop and used as a 'starter'. Products of the latter type should be used in larger quantities to begin with, say, 4 tablespoonfuls per pint (0.6 l).

Other Products

After milk, I suppose cheese and butter are the first products to come to mind, but there are others.

If you are interested in making sweets you will find that fudge and coconut ice are simply delicious and are always well received. Many health shops are only too glad to have such pure products to sell, and they also make excellent Christmas and birthday gifts. Any book of confectionery recipes explains how to make these sweets: simply replace cow products with goat ones.

The meat of young milk-fed kids is white and much like chicken, while that of older kids, say three months on, is more like veal. A stage further on than that and it tastes like lamb, but with more flavour; the liver is absolutely the best you will ever taste. Remember that kid meat is all lean and when cooked

should be wrapped in foil with some lard or margarine included to baste it; otherwise it might get a little dry. Meat from somewhat older animals can be cured like ham and is excellent dealt with in exactly the same way; old cookery books give methods of making goat bacon. You can also pickle goat meat in brine like salt beef, making it excellent in flavour and very tender. Castrated males weighing about 50 lb. (23 kg) at six months will give a dressed carcase weight of about 20–25 lb. (9–11 kg).

The skin can be cured and made into many useful and decorative articles such as slippers and gloves. Mounted on felt or thick blanket material, a whole kid skin makes a very attractive pram cover. If you ask the slaughterman to keep the skin for you he will do so, but charge you slightly more for doing the job, as he would have got some price on the hide.

If you decide to have a go at making produce to sell as well as use for your family, do please remember that you come under certain rules and regulations, not those governing ordinary milk, as would seem reasonable, but under the General Health and Hygiene legislation and 'Soft Drinks'. This may seem odd, but it is correct. If sold, goats' milk must be so labelled; it cannot just be called milk, as it seems that only cows are allowed to produce that. The Local Sanitary or Health Department will always advise you about these rules and regulations but will then come and test your goats and inspect the premises where the produce is packed, and it *must be outside your domestic premises*. So just to be safe, have some suitable clean place with wash-down walls and floor. All products sold through a shop must also be clearly labelled with your name and address.

When you agree to supply customers with your surplus milk, please remember that the winter yield from each goat will be only about half the summer's output. You should not take on customers in summer whom you cannot supply in winter; it is tiresome for everyone concerned. It is better to use a deep freeze and conserve the milk for your own use at the thin time. One advantage of goats' milk is that as it freezes more slowly than cows' milk, the smaller fat globules do not rise to the top and separate on thawing. Goats' milk freezes well and thaws correctly, unlike cows' milk. Cream will freeze only if you whip it and add a slight amount of icing sugar, otherwise it granulates.

Note

Recently it has become possible to obtain churns for holding milk in one-, two-, three-gallon (or equivalent kilogram) sizes, constructed in Dairythene which is much lighter in weight than aluminium, and will not rust or granulate when continuously cleaned with sterilisers. It is also possible to have an "inchurn" cooler, which fits into a two- or three-gallon churn. This is attached to a tap by a short length of hosepipe, through which cold water runs into the churn. The water then drips slowly down the exterior of the churn, thus cooling from the inside and the outside at the same time. The dripping water can be caught in a tank, or something similar, and the churn left standing in it when disconnected from the cooler.

Milk Recording

There are several good reasons for keeping records of your goats' yields. First, you can see just which animals are doing best under your methods; secondly, if a goat is sick her yield immediately goes down, so a sudden inexplicable drop should be viewed with suspicion and the cause for it should be sought. If some new food has been given the animal may be suffering from a digestive upset. If it is autumn she may well be coming into heat. All goats' yields decrease during this heat period and then pick up again once it is over, whether they have been served or not.

A goat's disappointing summer yield may be compensated by a relatively small winter decrease. Some goats' yields are high in summer, then drop right off in winter; others are more steady throughout the year. The latter type of longlactation goat may be the better bet because winter yield is very important. Demand for milk often increases at that time of year, when milk puddings become more popular.

To record the yield you can make yourself a chart or for a few pence buy a special book from the British Goat Society. The name of your goat is put in one column, with places alongside

CONTROLLED MILK RECORDS

ENTERED HERD	LEFT HERD	IDENTITY MARK	EAR MARK AZIE		P 1	NP 2	GU 3	CHANGE OF STATUS		FOLIO		
								NEW	DATE	3		

NAME OF ANIMAL	ALDERKARR KESSA				BREED BH 7154	HERD BOOK No. 273284	DATE OF BIRTH 1·3·70

BREEDER	MRS. L. V. HETHERINGTON		OWNER (if different)	
HILL FARM, MENDLESHAM GREEN, STOWMARKET		REGION SUFFOLK		REGION

NAME OF SIRE	844/31 LINWELL LARROMO		HERD BOOK No. 26963M	BREED HB	CHECKED BY
NAME OF DAM	RM33 ALDERKARR KIERON		HERD BOOK No. 266164	IDENTITY MARK	

DATE OF PREVIOUS CALVING / CALVING	DATE OF CALVING	LACT. No.	YIELD (lb)	LACT. DAYS	DAYS MILKED 3x 4x	% FAT	NO OF TESTS	TOTAL WT. OF FAT	LACTATION END			CHECKED BY
									REASON	DATE	CODE	
AGE AT 24 / INTERVAL / DAYS DRY BEFORE	6·3·72	1	2221 $\frac{14}{16}$	365		3·35	12		B·L·	10·3·72	Q / O USE 72	
AGE AT 36 / INTERVAL 388 / DAYS DRY BEFORE	29·3·73	2	2718	365		3·47	12		B·L·	24·74	Q / O USE 73	
AGE AT / INTERVAL / DAYS DRY BEFORE											CODE / O USE	
AGE AT / INTERVAL / DAYS DRY BEFORE											CODE / O USE	

Lactation Certificate

SEQ 8	GOAT HERD						MRS L U HETHERINGTON HILL FARM, MENDLESHAM GREEN, STOWMARKET, SUFFOLK.					REGION 3	

CONTROLLED MILK RECORDS — Weighing Date 17/18 June — MONTHLY STATEMENT — Period Ending Date 3/6/70

Herd Ref. No.

	9/12/69 RE ENTERED	28 3 68	4	26 3 70	5	RRNS63	WAYWARD ROSINA	2 0	on	loan	1	
		20 2 69	2			A220	ALDERKARR DAMARIS	2 12	3 10	6 6	2	4 1023 4/14 MASE
Nov 69 SOLD		1 3 68	1			A234A	ALDERKARR KIERON	4 0	5 1	9 9	3	1,1181 6/12 121
7 6 69 DI		19 6 69	1				SHANNOCK				4	
27 7 69 BT		7 3 69	2	16 3 70	3		ALDERKARR DIMITY	6 6	8 1	14 1	5	1,318 15/16 106
		2 3 70	1				ALDERKARR NYGELLA	2 9	3 12	6 5	6	853 8/16
		2 3 70	1				TIREE PENNY	2 9	on	loan	7	
		6 3 70	1				MELLIS KEKOMI	3 14	4 2	8 0	8	936 4/14
		20 3 70	1				ALDERKARR FEREDOT				9	
		31 3 70	1				ALDERKARR ESME	3 8	4 6	7 14	0	474 9/16 61
		31 3 70	1				ASHBECK SORCERESS	3 1	4 4	7 5	1	438 7/16 61
		10 4 70	2				ROSE BILBERRY	4 9	6 1	10 10	2	323 13/16

A monthly test record with cumulative totals on the right

for recording amounts, which until October 1977 was measured in pounds but is now measured in kilos. So an entry might read:

Mary a.m. 2.1 p.m. 1.2 total 3.3 kg.

You can weigh or measure the milk daily, weekly or even monthly, if you like.

On the specified day each month the milk is weighed and a sample taken from each milking. The milk recorder collects the sample, which will then be tested for its butter fats and protein content. Three copies of the record are made, of which one is kept at home, one goes to the BGS and the third to the MMB. At the end of 365 days from kidding, the yearly yield is given and also the average percentage of butter fats. Any goat which reaches 1,000 kg with an average of not less than 3 per cent butter fats gets from the BGS a prefix of R100. For any fifties up to the next thousand she gets a five—thus 1,050 kg would be recorded as R105, and so on. This is a great help when looking for good males, as you will know how much milk his dam has given. Should his sire's dam have been recorded he gets a prefix such as R120/154, where the sire's dam gave between 1,500 and 1,550 kg.

As the Milk Marketing Board recording fees have gone up considerably during the last few years, many more Goat Clubs are now using Club recording methods. The procedure is the same apart from the checking, which is carried out by several members. Spot checks are always made, as with the MMB, during the 24 hours following the actual date of recording. A neighbour, or some other willing person, watches you milk, then he or she weighs the milk, takes a sample from the pail after each goat is milked, places it in a numbered sample bottle and writes the amount against the goat's name on the Milk Sheet, morning and evening. The samples are then taken, or posted, to the Club Milk Recording officer, who hands them over to a MMB official. The resulting tests are the same as the MMB official tests, i.e. for butter fats and proteins. Each month's sample is detailed and sent back and the collective total carried forward until the end of 365 days from kidding. This gives the goat's yield in kilos. The BFS and protein averages over the year will also be given on a Lactation certificate.

There is at present a move afoot to try to get goat owners to agree to a 305 day lactation, as they do with cows, and with goats elsewhere in Europe. This suggestion is being resisted by goat keepers here who claim that one excellent point in favour of the goat is that she will milk through two years without kidding; this means that there are not so many surplus kids to deal with, and the milk throughout the winter is very useful.

Showing

This can be great fun, so long as you do not take it too seriously—especially if you are showing kids, as most people do when they start. Kids do change so fast; they can look good and come top one week; then two weeks later, losing their baby coat, they can look terrible, only to improve again once moulted out.

All goat clubs and many county agricultural shows have goat shows yearly, so join a club close to you and watch their way of doing things. You will see how to make the kid walk nicely on the lead and how to make it stand well and to show it. Some kids love all the fuss and really show off; others hate it and sulk.

The show schedule—which you should be careful to read and fill in correctly—will usually have the following abbreviations:

S & BS	(Saanen and British Saanen)
T & BT	(Toggenburg and British Toggenburg)
BA	(British Alpine)
AN	(Anglo-Nubian)
AOV	(Any Other Variety)

The scope of AOV frequently worries people—well, it means anything unspecified, i.e. British (HB), Golden Guernsey (GG), Foundation Book (FB), Supplementary Register (SR) and Identification Register (IR). The initials of your goat's breed will be found at the top righthand corner of her registration card.

If you are going to small shows—afternoon competitions—always take a few branches and some hay in a net or tied in a bundle. Each animal will need a collar and lead as well as a good cord with which to tie it to the post. Young kids will also need a bottle of milk and a teat.

In bigger shows with milking competitions, pens are provided, made of hurdles. If you are entering a milking competition you must arrive the day before the show and stop overnight. You have to milk your goats by a certain time and then the stewards will strip out all the goats, so that all of them start equal.

The competitions have a complicated points system involving:

- For each calendar month since kidding (1/10 of a point is given, reaching a maximum at one year of 1.2, after which no more points are added);
- pounds of milk produced (one point per pound);
- amount of butter fats (five points for each four ounces).

To get eighteen points and a coveted milking star, the goat must give at least a gallon (4.5 l) in twenty-four hours and the milk must contain at least 3.25 per cent butter fats at each milking. If twenty points or more are awarded

All show goats must be neat and well presented

and this proportion reaches 4 per cent at each milking, the goat then receives Q*. With this points system you need a calculator to work out the result—and indeed it is some time after the show when all the analyses are completed, the points declared and the milking awards made.

Competitions are always a chancy business, because travelling upsets some goats. But all exhibits should be sponged over first with warm water, then with a soapy liquid (not a deter-gent, it makes whites appear cream). Although the goat will not love you for this attention and will probably leap about, work up a good lather all over the animal, taking care to mind its eyes. Rinse it well with warm water and dry with towels. If it is warm and sunny, let the goat run outside to finish drying; if it is chilly, tie a dry towel or coat over its back to keep it warm and make its coat dry flat. When you are at the show give it a brush and rub over with a piece of velvet for a good shine.

General Care of Goats

No matter how hard you work at the show, no goat which is underfed or wormy will ever look well. Among other factors maintaining a goat in good condition are:

- regular worming (see page 57)
- hoof trimming so the animal can stand or walk well (see page 57)
- good feeding.

Try to teach young goats to lead when they are about a month old; once taught, this will never be forgotten. If you direct it towards a pasture or its bottle, it will soon connect walking with something nice.

Do not ever let male kids play with you—especially butting your hands or legs. Although this may be acceptable in a small goat, by the time it is fully grown the same tricks may be dangerous, or at least will leave you flat on your back in the mud.

I have heard it said that goats are worn out at eight years and should be culled at that age. However, if you 'run them through'—kid them every other year—they will have kidded only four times. There should be some very special reason if a goat from any breed but Anglo-Nubian is kidded two years running. Anglo-Nubians are an exception because they are shorter-lactation animals—in their own desert countries they would kid twice yearly.

I have eleven- and sixteen-year-old goats in my herd which are in perfect condition, and I would never cull merely because of age—it is condition which matters. In her last lactation my sixteen-year-old gave 3,114 lb. (1412 kg). (I think this may be a record for her age; of course, she gave a lot more when young.)

Goats' teeth do deteriorate with age. They only have a set in the lower jaw; in the upper part of the mouth is a large pad against which food is held for eating. When a kid is born its teeth in the centre front are like a child's milk teeth. They come out and are replaced by mature teeth at the rate of two a year, a 'full mouth' of eight teeth being reached when they are four years old. From then on they wear gently down—though not so fast as sheep's teeth.

Keeping a Male

There is a quick answer to that, *do not* —at least, not a male for stud.

I know this will sound hard when you are confronted by a small, attractive kid—but in six months he will be far less pleasant. The rutting season lasts for five months, and during that time males develop a most pervasive odour which adheres to whatever it touches and is very hard to remove. This means you must wear some protective clothing, such as a full-length plastic coat, for everything you do for the male. You must also keep him well away from the females of all ages, and particularly from your dairy area. This smell attracts 'his ladies' and repels all others. He will cost money to feed all year without

Trimming hooves
Tie up the animal or get someone to hold it. To trim front feet, stand back towards the goat's head and trim the flap edge level with the sole. The heel—the pad at the back of the foot—should be trimmed level with the rest, and it should all be finished off with a few rubs of a Surform file. Do not try and remove all the edge at once, otherwise you might cut too deeply. Do a little and look at the work—if it is pink, stop. To trim back feet, you should again stand with your back to the animal. Draw its foot up between your knees so that you can hold it firmly in a position which is reasonably comfortable for the beast. The best tool for trimming is a pair of footrot shears. However, these are expensive and garden secateurs will do nearly as well. Knives such as a Stanley knife or lamb-hoof knife with a curved tip can also be used—but always remember to slice away from you so that you run no risk of slashing yourself

giving milk in return, and he will need special housing, separate from the others. It is cheaper and less trouble to use a good male owned by someone else. If males are to be kept for meat they may be castrated by the Ring method. This is a special rubber ring put with a special pair of forceps (shepherd or calf-rearer) round the scrotum of the kid before he is one week old. The male odour will not then occur and he will be unable to serve a female. How-

ever, it is an expensive luxury to keep an animal which cannot 'work' for his living. The quantity of food eaten by the male will feed an extra milker with much greater rewards. Male goats do not put on much fat and are not really an economical proposition kept for meat except for those who literally have milk to waste.

As I have said before, you will be able to get information on the nearest suitable male for stud from the annual lists of the British Goat Society or from your local goat club.

Unless a male kid is from an exceptionally good milker which is also of excellent conformation, and sired by the son of an equally good dam, he is unlikely to be kept. Most come from herds where milk recording is done by the Milk Marketing Board and where the goats have prizes to their credit at agricultural shows. Your choice of sire makes or mars your herd for years, so find and use the best. In fact, it may well be worth finding out what males from each breed are in your area before you decide what breed of goat to purchase.

Management

Use your time, money and available space wisely; think first and cut out endless journeys. Assemble in your goat house, or in close proximity to it, as much as possible of food, hay, straw, water, utensils—anything you will use. Order your hay before you get short. Both hay and straw are cheaper bought off the field, so if you can store them dry buy the whole year's supply at one time.

Make yourself a timetable, and stick to it. Feeding and milking should be at fixed times, uneven milking times causing an animal to dry off. Apart from the overnight one, kids' bottles need to be given at the same intervals; and they should be kept as clean as an infant's.

Hooves should be trimmed once monthly; an animal is only as good as its feet.

Worming should be done spring and autumn. A mated goat should be wormed seven weeks after service and three weeks after kidding. One 5 g tablet of Thibenzole for every 100 lb

Secateurs, Stanley knife of footrot shears can all be used for hoof trimming

(45 kg) is used, goatlings or kids needing half or quarter according to weight. Tablets may be pulverised, mixed in water, and given as a drench (see page 61); kids can have theirs in their bottle of milk.

The British Goat Society will not accept for registration any kid which has not been tattooed in its right ear. This type of forcep, which has a self-release spring, presses the letters (prefix, number and year letter) instantly into the kid's ear, which should then be rubbed with tattooing paste. Your local goat club or the BGS will issue you with number and letters.

Suggested timetable for summer and winter

07.15 hrs: feed kids with milk (warm); give older stock concentrates.

07.45 hrs: remove feed pails; bring out goats one at a time. Milk in usual order; return goats to pens; place milk to cool. Fill up water pails for each animal; fill hay rack.

10.00 hrs: (*summer*) let all stock out on to grass, if dry, for grazing; let yarded animals into yard; feed green food in racks.

(*winter*) keep stock in unless dry and fine.

12.00 hrs: (*summer*) bring in young kids and shut up for half an hour before feeding milk; release again half an hour after feed.

(*winter*) feed kale or green food to goats kept in; refill hay racks.

16.00 hrs: bring in all stock; pen and leave to cud.

17.00 hrs: feed second half of concentrates; leave until 18.00 hrs, then repeat morning routine; kids' feeds must be fitted into your timetable to suit your convenience; kids' last bottle should be given after evening milking, making a shorter interval, as the night is a long gap for a very young animal.

This is simply an idea of what is involved; each person does it in his or her own way. Children often like feeding the kids, and can do it before and after school. This keeps them interested, and they may also like to collect wild green as well.

Do not plunge too deeply into goat-keeping until you know you like the animals. Once in, they become part of the family and grow in number unless you are very strong-minded. Your herd multiplies very quickly when kids arrive in twins, triplets or, sometimes, quads.

Small kids are not toys, and should not be played with so long that they become exhausted. And, as I have said, neither should they be allowed little tricks, such as jumping on your back.

Other Kinds of Stock

Most young stock do very well when reared on goats' milk. Calves, lambs, puppies and kittens all thrive on it, and it is also good for fattening up poultry and rabbits.

You can buy a Channel Island × Angus calf when it is about a week old, as I do every year. When you get it home, place it in a dry shed, with plenty of straw round it. Boil water and for the first twelve hours give three feeds of 2 oz. (57 g) glucose in 2 pints (1.1 l) of the water cooled to just blood heat. On the second day make a mixture of half goats' milk and half glucose water; on the third day give 2 pints (1.1 l) of goats' milk three times during the day at equal intervals. From then on, give 3 pints (1.7 l) night and morning, with some soft hay in a rack for it to investigate. The amount of milk can go up to 4 pints (2.3 l) night and morning, and be kept at that for ten to twelve weeks. From the second week, give a little of the food you are giving to the goats— any nice meal will do. If you really wish, you can buy special calf weaner nuts, but the calf will do very well on what you have. Place a handful of meal into the milk pail when it has finished sucking up the milk. Gradually it will eat more; then omit one milk feed, but do remember to place a pail of water near to where the animal will eat. There is no need to let the calf out of its shed; it is warmer inside during cold weather, and nothing will be gained by its grazing.

I keep my own calves until eighteen weeks, by which time they weigh about 300 lb. (136 kg). They are then taken to the butcher to be killed and dressed. That gives my family about 125–150 lb. (57–68 kg) of boned meat, which is tender, but not chewy like very young veal. It works out very well in price. I cost in price of calf, collection, milk, meal, hay and straw, removal to abattoir, cutting up and packing, and my collecting the meat. Overall it costs about 27p per lb. in 1977—what can you get for that in the shops? Really, it is a very good outlet for your extra milk.

Lambs need to be bottle fed; they are usually very young when you get them. Frequently they are orphans, or else triplets with which the dam cannot cope and the shepherd has not the time or milk to do so.

These tiny creatures probably have not had the so essential colostrum. That can be imitated by adding 1 teaspoonful of liquid paraffin and 1 egg to 1 tablespoonful of glucose in $\frac{1}{2}$ pint (0.3 l) of milk. Whisk it all together, and offer the mixture to the lamb in a small bottle fitted with a plug-in lamb teat as used for kids. The liquid should be just warm, and given in amounts of about 4–8 fl. oz. (114–227 ml) at a time. After the first day warm milk alone will be in order. Lambs do not need so much milk as kids, nor for so long. They should be given no more than 3 pints (1.7 l) a day, divided into three or four feeds to begin with, reduced to one night and morning when the lambs begin to graze freely. They will, of course, graze, not browse, like kids. Do not leave young lambs out at night: they have no dam to tell them where to get out of the wind, and are better in a shed.

About two months on milk will give them a good start. After a couple of weeks they will appreciate a little meal and hay, and make much better carcases for it. Do not stint them—after all, they are going to feed your own family a little later on.

Disposal of Kids

If you are disposing of unwanted kids, I would strongly recommend you not to put them on to a market. There is a good deal of young stock bought by various sections of the population whose methods of killing are not such that one really cares to dwell on it. Most of us would rather have the kid painlessly put to sleep than subject it to such an ordeal.

If you decide to dispose of kids at birth, or at four days when taken from their dam, chloroform is the kindest method. The chemist will supply you with a small amount if you tell him your reason for needing it. Collect a jar with a wide neck and put into it some cotton wool dampened with chloroform. Place the kid's nose lightly into the neck of the jar; it should not be pressed in hard so that it becomes frightened. Soon it will fall asleep. Add more chloroform and then push the nose firmly in and hold it there for a few minutes, when it will be ready for disposal.

Some veterinary surgeons will supply you with Nembutal or a similar tablet. This can be placed in milk and, when dissolved, fed to the kid; it will not wake up.

This sounds dreadful, but killing the kid is really kinder than giving it away as a 'pet'. Many such kids have a frightful life. People do not realise that kids need milk-feeding for a long time; the poor little creatures are tethered out in all weathers and expected to find their living in the hedge bottom when they are far too young for such treatment. Often this is the result of pure ignorance. It is especially wrong to let uncastrated males go to such 'pet lovers'. When the male begins to smell —as he must in the autumn—he will soon lose favour and end on the market.

A number of poulterers will buy quite young kids once they weigh about 10 lb. (4.5 kg). These are for immediate slaughter and are properly dealt with on licensed premises. It does seem foolish to bury a perfectly good kid when it could be used to help someone to feed a family, so I would rather it went to a poulterer. Fortunately, more people are rearing their male kids for their own use. Freezers make it possible to make better use of the meat, so that the family does not have to use a whole kid carcase at one time; previously that is what had to be considered.

Some abattoirs will buy carcases weighing out at 25–30 lbs giving 50–60p per lb, for well-fed kids of about three months.

Female kids are always easy to sell. An advertisement in your local paper will bring quite a surprising number of enquiries, especially if you can say 'Pedigree Female Kid, Reg.', then its age and breed. The possible purchaser has some idea of what to except, and if you talk to other goatkeepers you will know roughly what price to ask.

Some Do's and Don't's

Do   remember to give fresh water at least twice daily to all animals; no animal can produce milk or young without water.

Do   remember always to feed hay to all goats before turning them out on to grass or other grazing, at all times of the year; it goes a long way towards preventing entero-toxemia and bloat.

Do   remember to ease the tight full side of the udder of a newly kidded goat; no matter how many kids there are, they usually suck from one side only. If you do not keep the unsuckled side soft, the mother may well develop mastitis.

Do   Remember to trim goats' hooves monthly. This means trimming off the flap which grows round the hoof; if left, mud, stones, etc., can lodge underneath and cause damage to the foot, resulting in lameness.

Do   remember to join either the British Goat Society, or your local goat club.

Don't ever keep horned and hornless animals together, particularly in their house.

Don't feed frozen greenstuff; at best it will cause scouring and digestive upset, at worst aborted kids.

Don't leave a goat tied up, tethered or otherwise, when she is due to kid; she, or her kids, can get tangled in the chain.

Don't leave your goats out in the rain or very cold winds—chills result; conversely, they can also get heat stroke from hot sun.

Don't let anyone's dogs chase your goats; the result can be aborted kids, bruised udders and mastitis, or even torn throats.

60

Ailments

General Precautions

Always remember to have the vet's telephone number at hand, and if no phone is in the house, at least one 5p piece available. Also remember that a little blood goes a long way. A cut may not be so bad as you think, as you will discover once you have applied pressure to stop bleeding and washed away some of the evidence.

A goat's normal temperature is 102°F (about 39°C). To take the temperature, the thermometer is rubbed with Vaseline or other lubricant and inserted into the rectum.

Your home medicine box should contain:

Thermometer
Drenching bottle
Bicarbonate of soda
Antiseptic such as Dettol
Antiseptic powder
Liquid paraffin (cooking oil will do)
Glucose
Iodine
Thibenzole powder
Sulphadimadine 5 g tablets
Udder cream or Vaseline
Adhesive tape
Large wad of cotton wool
Clean sheeting or crepe bandage
Scissors
Torch.

Drenching

Several treatments require drenching— that is, giving a liquid medicine. To do this, hold your left arm around the goat's neck, place your left hand beneath its chin and tilt the head slightly upwards. In the right hand hold a small long-necked bottle of the soda-water type. Holding the mixture, insert the mouth of the bottle into the side of the goat's mouth, and slowly trickle in the medicine. Make sure the goat is swallowing by watching its throat. If the animal coughs, stop instantly, or you may allow fluid to enter the windpipe and so into the lungs, which would cause pneumonia.

Drenching

Abortion

This is not in fact an ailment so much as a disaster. Since goats do not have *Brucella abortus*, there are two likely causes: first, the abortion may be mechanical, the result of a bump by another animal, or of slipping in mud and injuring herself; secondly, it may result from eating frosted green vegetables. The animal should be treated with quietness and light feeding as if it had kidded; so long as the animal has cleaned correctly—that is, dropped the placenta—she will be all right. If the goat is far enough in kid to have developed an udder, commence milking to increase the yield.

Abscesses

Sometimes these come up on jaw bones, the result of pricks by thorns; they take a very long time to point, but can be encouraged to a head by using white oils gently dabbed on the area. Once they are pointing, lance them with a sharp blade; wash out with warm water and disinfectant, or peroxide, and bathe the places for a few days until quite clean.

Acetonemia

A production disease, usually contracted shortly after kidding. The animal refuses her concentrates, is miserable and restless, and smells of hay or nail varnish. Treatment: the animal should be drenched once daily with 4 fl. oz. (114 ml) glycerine or Ketol in 9 fl. oz. (256 ml) warm water; if there is no rapid improvement seek veterinary assistance. Glycerine can sometimes be given as a drink. Cortisone is usually given; black treacle or molasses is also very good for treating this condition. As with glycerine, give as a drink or drench.

Anaemia

The goat becomes progressively less interested in food; its condition goes down and it grows listless and apathetic. This is often caused by heavy infestation of worms and lack of cobalt. Treat the worm condition with Thibenzole; give cobalt salt in food, and have a cobalt lick always available. Seaweed meal will help, added to food or offered in a bowl in limited amounts. About 12.5 g daily of seaweed meal and 20 ml Parrishes' Food sprinkled on the concentrates, is very efficacious.

Arthritis

This can be caused by the animal lying on cold, hard surfaces in early life. The joints affected stiffen and swell, the animal going on its knees rather than walking. To relieve the pain crush six aspirin tablets and give them in a drench of warm water. There are some horse powders which are also effective for goats. These can be obtained from your veterinary surgeon and vary in name with the maker. Though arthritis can be retarded, the condition will usually return with age. It should be prevented by ensuring that animals have thick, dry straw beds free of ground draught.

Blindness

Temporary attacks may be caused by the goat banging its head on hay racks or similar objects. These can be assisted by bathing the eye and applying soothing lotion of bicarbonate of soda. Longer-lasting attacks are usually caused by New Forest Disease, which appears to be fly-borne. This causes the eye to film over so that it looks grey; it runs freely, is obviously painful and oozes puss. Bathe the eye, putting in Chloromycetin eye ointment or eye drops, and generally easing the pain. In some cases antibiotics are needed.

The condition will usually clear in about two weeks, leaving a blurred eye which does seem to have sight in it. Once a goat has had this condition, it has similar attacks in other years. The disease does not appear to be infectious.

Bloat

This is not an illness but a condition which can kill. Too much wet grass or lush green food causes gas in the rumen (first stomach) to press against the exit. This causes distension as the gas is unable to escape. The animal swells and is in great distress, its body tight and hard. Prepare a mixture of 2 fl. oz. (57 ml) liquid paraffin and 0.4 oz (12 g) of bicarbonate of soda. Drench this into the goat and walk the animal about and rub its sides to relieve the gas. One hour later another tablespoonful of dry bicarbonate of soda may be placed directly on the animal's tongue. If, by a further half hour, no sign of improvement is seen in the form of gas being brought up, sides being less taut, dung being passed, etc., get veterinary help quickly.

Caprine Arthritis and Encephalitis

CAE is a very unpleasant ailment, caused by a virus, which very rarely shows up in animals under one year of age. It was not thought to exist in the British Isles until recently although it is of epidemic proportions in America and has a strong hold in many other countries. When it does appear in kids, it takes the form of weakness or paralysis of the legs. They are unable to stand or walk although there is no loss of appetite and they still appear quite bright. The only solution is to put it down. In older goats the disease may take the form of infection of lungs, joints or nervous system, udder lesions and hardness which causes falling yield, chronic pneumonia, or acute or slow failure of one or more joints. There is not necessarily any temperature rise or fall and the goat retains her good appetite. Blood tests will show whether the virus is present and, if positive, the animal should be put down as there is no cure. Transmission of the disease is mainly via infected colostrum and milk. Do not, therefore, whilst at a show, for example, feed kids with milk from an unknown source. Blood tests of all stock over one year old will show if an animal carries the virus. If an inkid doe is found to be infected, the kid should be removed from its dam immediately at birth without being licked or suckled; the mother's colostrum should be sterilised or discarded altogether and that of another animal used. The kid will not be affected as the disease is not passed on in utero, but the dam should be put down as she is a source of infection to other goats (not, however, to human beings). This disease can be stamped out now if firm steps are taken before it becomes prevalent.

Coccidiosis

Usually collected from infected grazing, this can be ingested after poultry have been allowed access to the land. Treat with Sulphadimadine tablets, giving one 5 g tablet once daily for five days to adults, but less to young stock.

Colic

This is a very painful condition in which the goat suddenly refuses food, collapses and convulses. Give 4 fl. oz. (114 g) of oil, with aspirin or other pain-killing tablets crushed in. Follow, when some bowel movement in the goat is observed, with glucose or spirits in 8 oz. (227 g) warm water or milk. Remember not to give this until dung has been passed.

Cuts

If wide open, wash the cut carefully, gather the edges together and hold with adhesive plaster. If it is bleeding freely, apply a wet pad and press firmly. When the bleeding has stopped, apply antiseptic powder and cover with a fresh pad. If the cut is deep, wash with peroxide of hydrogen, and cover the wound. In either case, if the animal has not been vaccinated against tetanus, have it done at once.

Entero-toxemia

Quite the worst sickness found amongst goats in Britain; its onset is rapid and often fatal. The animal scours, staggers and usually walks backwards. Obviously ill and in pain, it will go into a coma and convulse; death rapidly follows. It is easier to prevent than cure, and is caused by toxins contained within the goat, which some unforeseen set of circumstances triggers off. Vaccinate against the sickness; the vaccine also covers tetanus. Once treated, should the animal become ill with suspected entero-toxemia, give two 5 g tablets of Sulphadimadine, crushed in some milk, followed two hours later with six aspirins and a tablespoonful of Kaolin powder, also in a drench. For four days repeat the full treatment but with one Sulphadimadine tablet instead of two, every twelve hours. Feed no green food. The goat can have hay, water and, when it wants to eat, bran and soft food for a few days.

Fading

Fading is a metabolic condition in which goats will eat progressively less concentrate, get thinner, refuse green food and then hay. Cortisone injections and steroids are useful, but until the animal can be given very hard, dry herbage in the form of clover or lucerne hay, dry branches, etc., it will go on a hunger strike.

Once eating really hard food with a great deal of bulk, it will respond very fast. The animal will never do well in the area where this takes place and should be passed on, or put down.

Fluke

Fluke is carried by a small mud snail and is usually found only where pastures are very wet marsh grazing, and where sheep or rabbits are to be found. Symptoms are a dry cough, swelling on the brisket (chest) and under the jaws, slight scour and listlessness. The treatment is Hexachlorethane and Vitamin A as directed by the vet or the label of the drug used.

Footrot

Infection sets in when mud gets under the flap at the edge of the hoof. The bacteria may also infect the sole and part may have to be cut away. In less severe cases, an aerosol of Chloromycetin and gentian violet may be used —one is made specifically for this purpose. Hooves must be trimmed back and the infected material removed; washing in potassium permanganate will help to clean the area. Keep hooves dry and trimmed all the year to prevent the condition; walking on hard-surfaced yards helps to keep feet in a reasonable state.

Fractures

If the limb is swollen, but looks longer than usual, try and fix the two ends of the bone, pad round with cotton wool and strap into place with adhesive bandage over a splint of light wood or folded cardboard. If very swollen and not reducing quickly after applying cold water bandages, suspect compound fracture and get veterinary assistance, which should also be sought if the injury is close to a joint, where splinting can stiffen permanently. Keep the limb bound up over six to eight weeks, as the animals are very active and so need longer to heal the bone.

Grass Tetany

This is also known as Staggers, and is correctly called Hypermagnesimia. Animals with staggers often become highly nervous so that they scamper away from you and drop dead; alternatively they may become apathetic and disinclined to eat or move. Lack of magnesium in spring grass is the cause, and magnesium bullets can be given to animals in an area which is known to be deficient. Calcined magnesite may be applied to the grazing area and high magnesium licks and mineral powder should be available to stock, in the early part of the year especially. It will then be taken into the system prior to turning out on to grass, and prevent trouble.

Laminitis

The feet of the animal become hot and swollen, causing it great pain and tenderness; the whole hoof can slough off in bad cases, but rapid treatment will reduce the condition. This can be caused by too much protein in the diet, hard floors and too hot beds—that is, beds left too long without clearing out, so that they steam upwards. Soak each affected foot in warm water with permanganate of potash, then apply hot bran poultice; keep in place by enclosing in a polythene bag, covered by a sock bandaged on firmly. Replace every few hours until swelling is reduced. Remove protein from diet; give plenty of green food and meadow hay, bran, few oats, no barley. Keep the hoof well trimmed. The horn will be very hard, but poultice will soften it greatly.

Lice

Animals in poor conditions or dirty pens sometimes get lice. They can be easily cleared by use of a good proprietary Louse Powder sprinkled around the tail and shoulders and along the back. Males at stud may get these pests from visiting females and should always be given a good sprinkle and shampoo at the end of the rutting season.

Lumpy Jaw

Goats which graze gorse sometimes get sores round the mouth with swollen lips and swelling under the chin. It is amongst the deficiency conditions, the result of iodine shortage in the land. Paint sores with iodine and add 2 ml Lugol's solution to the diet twice daily. Penicillin injections are useful. Feed sloppy diet of mashes. Cover any abrasion on your hands and wash thoroughly after handling an infected animal.

Mastitis

This comes in several types. If it is *acute*,

the animal's udder becomes swollen, lumpy and very inflamed; the milk has clots and blood in it. The *sub-acute* condition is often hard to detect: the only signs are that occasional small clots appear and the milk will not boil, but the animal appears well. The *black garget* type causes loss of the udder from gangrene, and is often fatal. Treatment of the acute form needs veterinary assistance rapidly, probably in the form of injection with a quick-acting antibiotic, followed by a long-acting type. The less acute form may be dealt with by the owner: obtain Intermammary Penicillin tubes, and insert the contents of one tube into each side of the udder; milk out the infected udder after twelve hours, disinfecting the milk and disposing of it—it should not be used. Repeat the insertion of antibiotic and continue to remove and dispose of milk, which should not be used until 72 hours after last insertion; this should clear up the condition. Milk the animal last of the herd and wash hands thoroughly.

Black garget is rarely treatable. If with *rapid veterinary treatment* the animal survives, half the udder will probably slough off; the remaining half will then heal.

Metritis

This is infection of the uterus, usually following an assisted kidding or birth of a kid which has died *in utero*. The goat needs an antibiotic injection to kill the infection. The unplesant-smelling discharge, which is strawberry coloured, should be syringed out twice daily with a mild disinfectant, until it clears. A rubber tube and a funnel can be used for washing out the vulva if no syringe is at hand.

Milk Fever

Signs of milk fever are collapse of a newly kidded animal, convulsions and coma. It is caused by the milk rising too rapidly in quantity and depleting the whole system of calcium. Borogluconate injection under the skin has a rapid effect and will save the animal. A vet is required for this. It should be prevented by not giving high-protein diet immediately before or after kidding, and including steamed bone flour in small quantities (12 g [0.4 oz.] daily) to the diet prior to kidding. Bran is a help in the diet as it is balanced, but will not cause a flush of milk.

Pink Milk

Often found in first kidders, this is caused by small capillary vessels breaking as the udder distends rapidly with milk. Calcium tablets with Vitamin D will help to clear the condition; milk very gently for a time.

Pneumonia

The animal is miserable, refuses food, has cold nose, ears and muzzle, and is sometimes found to have sharp, rapid breathing and a harsh cough. Young kids choke and refuse milk, because breathing fast makes sucking impossible. Give the animal warmth and rugs, and if it is down prop it up on its side with sacks of straw. Antibiotics are needed for quick recovery. Sulpha drugs help greatly; Sulphamezathine can be given orally if dissolved in milk, but be cautious as breathing difficulties make drenching difficult. Give plenty of fluids with glucose. Food should be light—bran, oats, and green when the animal will take it, as well as good hay.

Although the disease is often a virus condition, cold wet conditions outside after a too hot stable will cause one type. Fresh air, no draughts, warmth and fluids are a must.

Poisons

Try and find out what has caused the illness and treat accordingly. Oil is usually useful in clearing the system, but you must look carefully at the pasture or yard and buildings, and find the source; it could be paint, weed-killer or plants. Treat for material causing the trouble.

Pulpy Kidney

As with entero-toxemia, the signs are foul scour, convulsions, coma and death. Prevent by vaccination. Various vaccines available cover also tetanus, braxy, black dysentery, etc. Vaccine injected in the loose skin over the shoulder, at one month and once yearly afterwards, will prevent these, as well as entero-toxemia.

Rheumatism

Stiffness on rising, obvious pain in one or more limbs, can result from cold, damp quarters for the young. The condition will deteriorate with age. Goats react well to aspirin drugs, and will live for a long time with rheumatism if kept without great pain. Goats are animals which do not take pain well; reduce pain and raise morale.

Rickets

Rickets appears in young stock, the affected kids getting bent legs, distended stomachs, distorted joints, puffy knees, etc. As it is caused by very badly balanced feeding with a shortage of Vitamins D and C, the animals need green food, correctly balanced feeding and sunlight.

Scurf

This is mainly found in older males which get insufficient exercise in the open air and access to grazing. Wash with mange solution and rub the skin with olive oil to remove scabby areas. Make sure the animal has a cobalt lick, which assists with hair and skin.

Tetanus

All stock should be vaccinated against tetanus, which otherwise is sometimes contracted when the animal has had a deep wound with bacteria upon it. If the stock is not vaccinated, treatment immediately after wounds is essential. Symptoms of lockjaw are the first things noticed: a very sick animal, which will arch its back, convulse and die in great pain if not treated quickly.

Glossary

AI Artificial insemination

Buckling Male goat between one and two years old.

Cloud Burst Name given to 'pseudo pregnancy' when nothing but water is carried by an apparently pregnant animal.

Drench Method of giving liquid medicine.

Goatling Female goat over one year and under two; or up to when she kids, if less than two years.

Kid Male or female goat up to one year of age.

Maiden Milker The phenomenon in goats when an unmated goatling develops an udder and produces milk. This must be drawn off or mastitis will result. The milk is normal and can be used in any way; frequently goes up to 3–4 lb. (1.4–1.8 kg) a day. Once the goatling has been served, she will gradually dry off.

Milker Goat of any age which has borne a kid.

MMB Milk Marketing Board

Service The act of mating.

Stud Male Any male, usually adult, who is at stud, 'is giving services'.

Sources of Further Information

The British Goat Society

The British Goat Society will accept members, family members, junior members or associates. The Secretary is:

> Mrs S. May
> British Goat Society
> Rougham, Bury St Edmunds, Suffolk.

The Secretary will also supply you with the name and address of your nearest goat club; there are more than fifty in the British Isles, Channel Isles and Northern Ireland.

Suppliers

Sectional wooden buildings from:

Browns of Wem	S & K Contractors
Four Lane Ends	Market Weston
Wem, Salop	Norfolk

Timber stables, etc., from:

Farmstead Developments Ltd	Flexinet from:
The Station	Bramley & Wellesley Ltd
Ashwell Thorpe	Gloucester Trading Estate
Norwich NR16 1EK	Hucclecote
	Gloucestershire GL3 4XD

Appliances from:

F. Ritson	The Gate House	D. Malpas Fabrications
Goat Appliance Works	Millstone Lane	31 Regent Street
Carlisle	Okerthorpe	Barwell
	Derbyshire	Leicester LE9 8GY
Harvester Goat Shop	Smallholdings Supplies	
Maylord Street	Priory Road	
Hereford	Wells	
	Somerset	

Foods and additives from:

Boots Farm Sales	Dunns Soil Fertility Ltd	Volac Ltd
Thane Road West	Corsham	Orwell
Nottingham	Wilts	Royston
(Mineral mixtures)	(Seeds, herbal mix etc.)	Herts
		(Kidolac milk feed)

Nutrichip Products
Gastard
Corsham
Wilts SN13 9QN
(Vitamin reinforced
food supplement)

Super Codlevine
St Cross
Beccles
Norfolk
(Mineral supplement)

Goat Nutrition Ltd
Biddendon
Ashford
Kent
(Caprilac kid milk replacer)

Herbal Products
The Hall
Kettlebaston
Ipswich
Suffolk
(Natural rearing aids)

Willow Tree Cottage
Biddenden
Ashford
Kent
(Caprivite, goat
nutriment)

Milking pails, churns, strainers, etc., from:
J. J. Blow Ltd
Old Field Works
Chesterfield

Milking machines for goats from:
Gascoigne, Gush & Dent Ltd
Berkly Avenue
Reading

Dairy thermometers from:
Astell Lab. Services Co
172 Brownhill Road
Catford, London SE6 2DL

Dairythene churns, coolers, etc from:
Lincs Smallholders
Thorpes Fendykes
Wainfleet
Lincs

Goat milk cartons from:
Bowater Perga
Princes Way Estate
Gateshead
(Wax cartions)

Lakeland Plastics Ltd
Alexandra Road
Windermere
Cumbria
(Plastic cartons)

Swains Packaging Ltd
Brook Road
Buckhurst Hill
Essex
(freezing bags and gauges)

Polypropylene cheese moulds from:
W. N. Boddington & Co Ltd
Horsmonden
Kent TN12 8AN

Cheese rennet from:
Fullwood & Bland Ltd
Ellesmere
Salop SP12 9OG

Christian Hanson Lab. Ltd
476 Basingstoke Road
Reading BG2 0QL

Lancrolin oil for curing skins, from:
Watkins & Doncastor
Four Throws
Hawkhurst
Kent

Index

abortion 61
abscesses 62
acetonemia 62
anaemia 62
Anglo-Nubian 12
arthritis 62

bedding 18, 19, 23
blindness 62
bloat 63
bran 28
breech birth 40
breed, varieties of 9–12, 53
breeding 34–36
British 12
British Alpine 9
British Goat Society 12, 13, 60
British Saanen 12
British Toggenburg 9
bulk foods 29
butter 46

caprine arthritis 63
care of goats 55
cheese 47
chloroform 59
clubs, goat 53
coccidiosis 63
colic 63
concentrates 26
cream, clotted 45
cuts 63

diet 28
disbudding 43
doors (of goat house) 19–20
drainage 18
drenching 61

earmarking 57
electric fencing 24
encephalitis 63
English 12
English Guernsey 12
entero-toxemia 63

fading 64
fattening stock 58
fencing 23
flooring 18
fluke 64
food 25–33
    analysis 27
    crops 30
    racks 21–22
    growing 29–30
footrot 64
fractures 64
fruit 25

gestation 35
goat house 16
goatling 13–14, 35
Golden Guernsey 12
grass 29–30
grass tetany 64
greens 25

hay 19, 22–23, 25, 30, 60
hoof trimming 54, 57, 60
horns 43
housing 16–20

insemination, artificial 36
insulation 18

kidding 37–40
    abnormal 39–40
kids 13
    disposal of 59
    feeding 41
    rearing 40
    sexing 40

laminitis 64
lice 65
lighting 20
lumpy jaw 65

males 55–56
management 57
manure 19
markets 13
mastitis 65

mating 34
meat 48
medicine 61
metritis 65
milk 45–49
    fever 65
    recording 50
milking 44
    competition 53–54
mineral licks 19

Nembutal 59

pails (food, water) 21
pens 20–21
pink milk 66
plants, poisonous 31–33
    wild 31–32
pneumonia 66
poisons 66
pulpy kidney 66

registration 57
rheumatism 66
rickets 66
roots 25
Saanen 12
scurf 66
showing 53
stud 34, 56

teeth 55
temperature 61
tetanus 67
tethering 24, 60
timetable 57, 58
Toggenburg 9

upgrading 36

veterinary surgeon 43, 61

water supply 16
wet weather 19
windows (of goat house) 20
worming 57

yards 19, 29
yoghurt 47

Toggenburg goatling, Alderkarr Phaedra, aged 13 months

Toggenburg buckling, Alderkarr Phaezzon, twin of Phaedra